U0315188

DEVELOPMENT AND PLANNING OF CHINESE CULTURAL AND COMMERCIAL STREETS
——MODERN RIVERSIDE SCENE AT QINGMING FESTIVAL

中国文化商业古街
开发与规划资料集
——现代清明上河图

金 盘 地 产 传 媒 有 限 公 司 策划
唐艺设计资讯集团有限公司 编

上

海峡出版发行集团 THE STRAITS PUBLISHING & DISTRIBUTING GROUP | 福建科学技术出版社 FUJIAN SCIENCE & TECHNOLOGY PUBLISHING HOUSE

图书在版编目 (CIP) 数据

中国文化商业古街开发与规划资料集：现代清明上河图：全三册 / 唐艺设计资讯集团有限公司编 .—福州：福建科学技术出版社，2018. 3

ISBN 978-7-5335-5565-8

Ⅰ. ①中… Ⅱ. ①唐… Ⅲ. ①商业街 - 城市规划 - 中国 Ⅳ. ① TU984.13

中国版本图书馆 CIP 数据核字（2018）第 037389 号

书　　名 中国文化商业古街开发与规划资料集——现代清明上河图（全三册）
编　　者 唐艺设计资讯集团有限公司
出版发行 海峡出版发行集团
福建科学技术出版社
社　　址 福州市东水路76号（邮编350001）
网　　址 www.fjstp.com
经　　销 福建新华发行（集团）有限责任公司
印　　刷 恒美印务（广州）有限公司
开　　本 700毫米×1000毫米　1/8
印　　张 101
图　　文 808码
版　　次 2018年3月第1版
印　　次 2018年3月第1次印刷
书　　号 ISBN 978-7-5335-5565-8
定　　价 888.00元

广欣商务宾馆
全国连锁
网吧
長沙油炸社
君鼎皇

Preface
序

Regional districts owning certain cultural deposits or traditional features are injected with cultural, leisurely and creative elements, so as to be renovated, extended or reconstructed as "international, cultural and fashionable" leisure commercial streets.

In recent years, after commercial real estates' rapid expansion, cultural commercial streets, with specific historical and cultural background, specific consumers and rich business activities, and combined with traditional culture, history and modern civilization, are popular in commercial real estate market. Their new development mode receives great popularity and is identified as a more intimate and sustainable mode.

With great popularity, domestic cultural commercial streets emerge in large numbers. On one hand, domestic commercial real estate magnates such as Wanda Group and Shui On Group all spare no expense to develop cultural commercial districts. On the other hand, unique historical and cultural deposits differentiate the cultural commercial districts from traditional commercial streets and form their own developing features and operation modes. In such a situation we edit this book themed with commercial street market positioning, planning and design, and operation, which includes many successful cultural commercial district projects on market, so as to provide all developers and designers interested in cultural commercial streets development with good examples for reference.

This publication includes three volumes with 44 representative cases collected from 30 cities of 18 provinces and municipalities in China. All cases are of high quality and wide scope, with high-definition pictures provided by our professional photographers on the spot.

According to development characteristics, the book is divided into 4 categories, i.e. commercial streets reconstructed from ancient residential areas and ancient streets, commercial streets extended from ancient towns, modern imitated commercial streets of ancient style, and commercial streets reconstructed on the basis of historic preservations, to present the development status and planning design features of domestic cultural commercial streets from all aspects.

There are 7 sections to introduce each case. From street background and market positioning, planning, design features, commercial activities and operation, to brand shops and cultural facilities, all detail descriptions are arranged for holistic presentation of the development and operation process of each case to provide general developers and designers with new and practical references.

文化商业街区，是指在有一定文化底蕴或传统特色的区域地段，注入“文化、休闲、创意”元素，改造、扩建或仿建出具有“国际性、文化性、时尚性”的休闲娱乐商业街区。

近年来，商业地产在经历同质化放量增长阶段后，文化商业街区这种以特定的历史文化为背景、以特定的消费人群为导向、以丰富的时尚业态为支撑、融合传统历史文化与现代文明的商业街区，受到商业地产市场的青睐。这种新型的商业地产开发模式受到业界的追捧，被业界认定为一种更具亲和力和可持续发展的经典模式。

在商业地产市场和业界的双重青睐下，国内文化商业街区开发日趋火热。一方面，国内商业地产大鳄如万达集团、瑞安集团、1912集团等纷纷斥巨资打造文化商业街区项目，商业地产市场上涌现出了一批成功的文化商业街区开发案例，如佛山岭南新天地、宽窄巷子、楚河汉街、上海新天地、南京1912等；另一方面，由于文化商业街区独特的历史文化底蕴，使文化商业街区的开发理念与开发模式有别于传统的商业街区，形成自己的开发特点、运营模式。在此种局势下，我们策划了此套以文化商业街区的定位、规划、设计、运营为主题的图书，收录市场上成功的文化商业街区开发项目，旨在为有志于文化商业街区开发的开发商与设计师提供参考和借鉴。

《中国文化商业古街开发与规划资料集》分上、中、下三册，共收录北京、天津、上海、重庆、山东、江苏、浙江、福建、广东、湖北、湖南、四川等18个省市30个城市的44个代表性项目，项目质量高、辐射范围广、数量大，项目图片全部由专业摄影师实地拍摄。

本书中，我们将项目按开发特点划分为四大类，分别是古住宅和古街道改建的商业街区、古城镇扩建的商业街区、现代仿建的商业街区和以文物保护为重点改建的商业街区，全方位展现我国文化商业街区的开发现状和规划设计特色。

在每个街区的编排中，我们从街区背景与定位、街区规划、设计特色、商业业态、市场运营、品牌商铺、文化设施七个方面详细介绍，力求全面呈现项目开发运营全过程，为广大开发商和设计师提供鲜活而极具实用性的文化商业街区参考案例。

CONTENTS 目录

古住宅和古街道改建的商业街区

Commercial Streets Reconstructed from Ancient Residential Areas and Ancient Streets

022-339

现代清明上河图

——忆清明上河图汴京古街之繁华盛景

北宋时，宋太祖赵匡胤定都于汴京，开宝元年『初修汴京，大其城址』，使汴京成为『四方所凑，天下之枢，可以临制四海』的重要都城。清明上河图就是反映汴京繁荣景象的浩翰画卷，是北宋风俗画作品，宽25.2厘米，长525厘米，是北宋画家张择端仅见的一幅精品画作。

清明上河图的整个画面以汴河为构图中心，按照景观变化分为三段，对北宋晚期汴京经济生活的各个方面作了详尽而生动的描绘。它以外城内东南角侧的农田园林为起点，向西沿着汴河溯流而上，经过内侧通津门外的上土桥、东角子门，到繁华的保康门街结束，描绘了有三四里之遥的旧城东南一带的繁华街景。

古住宅和古街道改建的商业街区

Commercial Streets Reconstructed from Ancient Residential Areas and Ancient Streets

022-339

上

福州三坊七巷 Sanfang Qixiang, Fuzhou 288
北京前门大街 Qianmen Street, Beijing 100
宁波南塘老街 Nantang Street, Ningbo 272
上海新天地 Xintiandi, Shanghai 074
黄山置地国际中心 Huangshan Land International Center 200
扬州东关街 Dongguan Street, Yangzhou 258

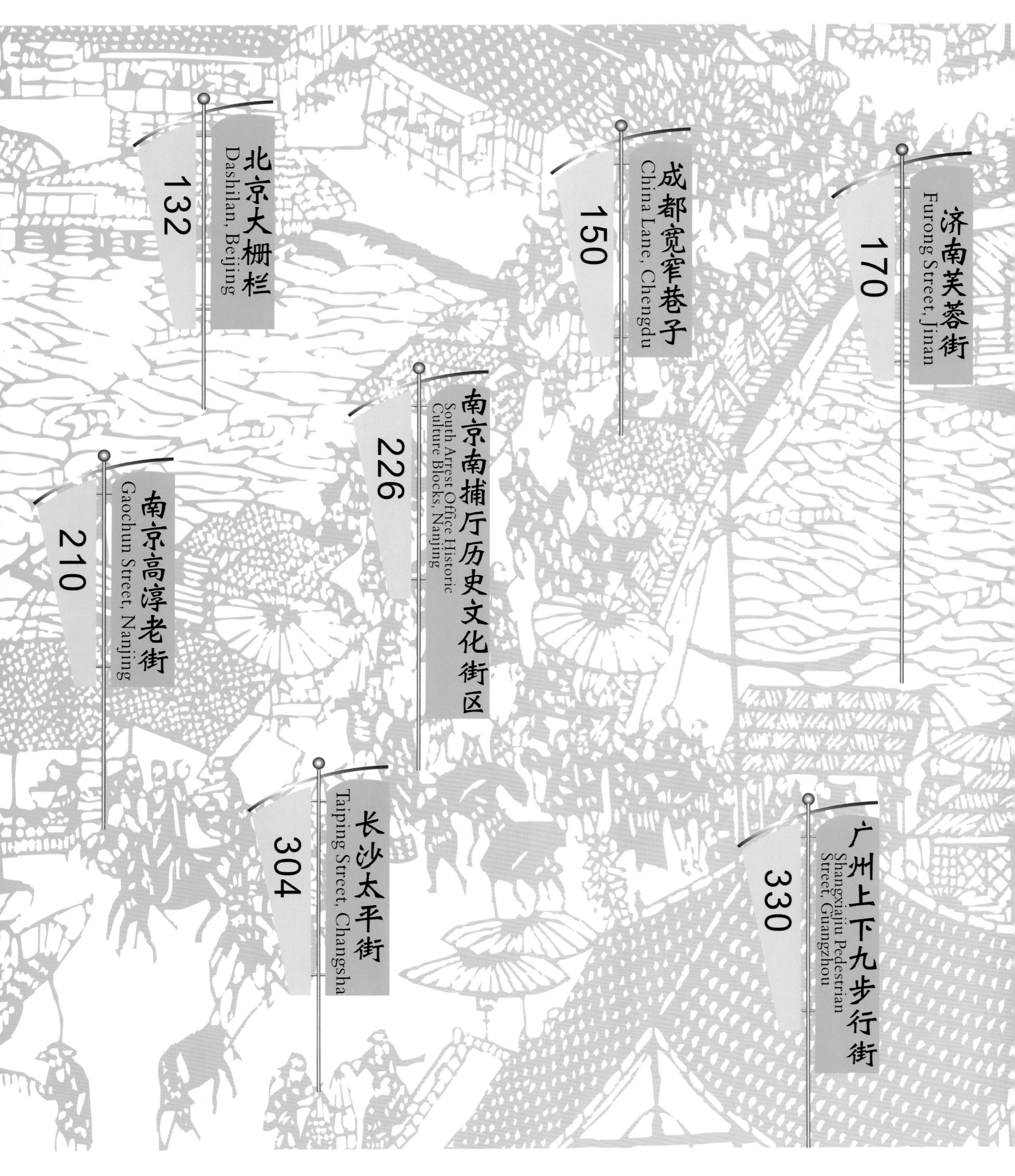

现代仿建的商业街区
Modern Imitated Commercial Streets of Ancient Style

022-262

中

078
上海城隍庙商业街
Chenghuang Temple Commercial Street, Shanghai
098
常州淹城传统商业街坊
Yancheng Chunqiu Tourist Area, Changzhou
154
开封宋都御街
Imperial Song Street, Kaifeng
166
张家界溪布街
Xibu Alley, Zhangjiajie
184
广州岭南印象园
Lingnan Impression, Guangzhou
246
北川商业步行街
Commercial Pedestrian Street, Beichuan
256
汶川中国羌城
China Qiangcheng, Wenchuan

古城镇扩建的商业街区

Commercial Streets Extended from Ancient Towns

022-129

下

044
上海七宝老街
Qibao Old Street, Shanghai
066
苏州周庄
Zhouzhuang, Suzhou
084
桐乡乌镇
Wuzhen, Tongxiang
098
大理南诏古街
Nanzhao Ancient Street, Dali
122
成都洛带博客小镇
Boke Town, Chengdu

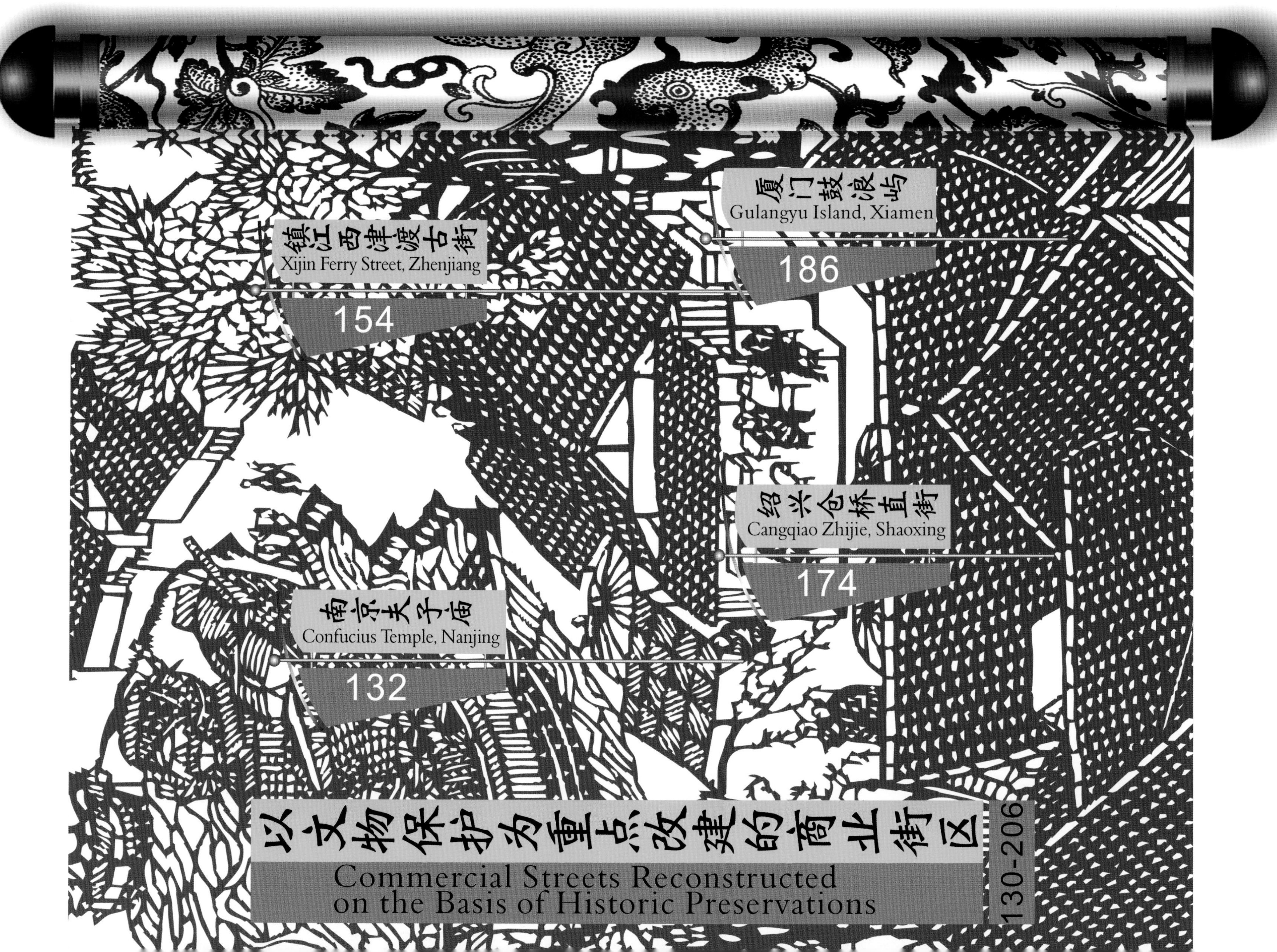
厦门鼓浪屿
Gulangyu Island, Xiamen
186
镇江西津渡古街
Xijin Ferry Street, Zhenjiang
154
绍兴仓桥直街
Cangqiao Zhijie, Shaoxing
174
南京夫子庙
Confucius Temple, Nanjing
132
以文物保护为重点改建的商业街区
Commercial Streets Reconstructed on the Basis of Historic Preservations
130-206

Modern Riverside Scene at Qingming Festival

In Memories of the Prosperous Bianjing

In Northern Song Dynasty, the emperor Zhao Kuangyin settled the capital in Bianjing. Later, the city was greatly expanded, and it was developed into the most important hub to govern the whole country. The famous painting *Riverside Scene at Qingming Festival* is just created to record the prosperity of that city. This masterpiece (25.2 cm X 525 cm) is the only work left by Zhang Zeduan, a famous painter in Northern Song Dynasty.

The painting composes picture on Bianhe River in three phases according to the landscape change and describes the economy and daily life of Bianjing in the later period of Northern Song Dynasty vividly from various aspects. The line of the painting starts from the farmlands and gardens at the southeast corner of outer city, goes along the river to the west, passes through the Shangtu Bridge and Dongjiaozi Gate, and finally stops

After homogenization stage of growth, the horrifying competition of commercial real estates makes the themed project be a focus that developers pay close attention to. Consequently real estates of themes of sport, tourism, culture and movie were born. Among them, development of culture-themed commercial streets gains popularity.

Commercial streets reconstructed from ancient residential areas and ancient streets can show the passed social features and historical connotations. Constructions lay emphasis on borrowing transmissibility and infection of history and culture, and convenience and modern sense of advanced facilities to create a new leisure experience mode which enables people to feel history when enjoying modern life. Features of planning, developing and design are as follow:

Feature one: relying on nature and selecting specific elements as cultural background

Concentrated historical building groups which have specific natural and cultural deposits are the focal areas of urban planning.

When developing and operating, historical, social and cultural elements, which can affect street value, are fully explored and specific elements which show local historical connotation best are selected as cultural background of the streets.

Feature two: having distinct consumption themes and abstemiously integrating with modern commercial elements

The consumers are fond of integration of historical culture and modern sense. Therefore, unique streets are built with the consumers' concern.

In order to satisfy consumption demands of specific group, modern commercial elements are abstemiously integrated and novel marketing mode is adopted in the space of combination of historical culture and modern business to provide consumers with different enjoyments in aspects of experience design, commodity display and activities.

Feature three: supported by abundant commercial activities and emphasizing relation between commercial benefit and street culture

Purposefully bring in fashionable commercial activities of consumer tendency, such as restaurants, bars, and stores in which visitors can consume with ease and enjoy nostalgic experience.

On the basis of bringing in international fashion elements, assembly development of local creativeness is supported to depress urban humanistic spirit and historical charm, and to create fashionable modern "city living rooms".

With the increasing influence of cultural commercial streets on regional economic development, the development of cultural streets is no more a simple building design or reconstruction. It has become a kind of protection, heritage and reuse of culture. Commercial streets reconstructed from ancient residential areas and ancient streets, through reminiscence of history and local culture, not only reveal local traditional culture, but also show modern fashionable life.

在经历了同质化增长阶段之后，商业地产竞争的惨烈使“主题化”成为开发商关注的焦点，体育地产、旅游地产、文化地产和影视地产等主题地产应运而生。其中，以历史文化为主题的商业街区开发备受青睐。

由具有历史或地域特色的居住区、古街道改造而成的商业街区，能够展现曾经的社会风貌和历史内涵，在凸显本土特色和商业发展中起着画龙点睛的作用。在街区改建中，侧重借助历史、地域文化的传播性、感染力以及先进设施的便捷性、现代感，创造一种新的休闲娱乐体验方式，让人们不仅能享受现代生活，更能感受历史、品味过去。在街区规划、开发、设计过程中，具有以下特点。

特点一：以原生态为依托，选择特定元素作为文化背景

街区本身或周边拥有集中的历史建筑群，具有特定的原生态文化底蕴，是城市规划的重点区域。

在规划时，关注文化带来的影响力。在开发经营时，充分挖掘能对街区价值产生影响的历史、社会和文化因素，选择最能表现地域历史积淀的特定元素作为商业街的文化背景。

特点二：具有鲜明的消费主题，有节制地融合现代商业元素

街区拥有喜爱历史文化与现代感交融的消费者，以他们的关注点为导向，打造出独具特色的商业文化。

为了满足特定人群的消费需求，有节制地融合现代商业元素，在历史文化和现代商业结合的空间里采取新颖的商业营销模式，从体验式设计、商品陈列、活动举办等方面给予消费者不同的享受。

特点三：以丰富的业态为支撑，强调商业利益与街区文化性的关系

有针对性地引入前沿、时尚、消费倾向明显的商业业态，如餐厅、酒吧、商店等，人们在这里轻松消费的同时，也完成了怀旧的历程。

在引入酒吧、艺术等国际时尚元素的基础上，支持本土创意力量的集合发展，通过具有创意性的购物、餐饮、娱乐、休闲、体育、会展、旅游等场所，浓缩城市人文精神和历史风采，打造时尚的现代“城市客厅”。

随着文化商业街区对带动区域经济发展作用的日益增大，它的开发已不再是一个单纯的建筑设计或者旧区改造问题，而是演变成了某种文化的保护、传承与再利用。由古街道、居住区改造而成的商业街，通过对历史或地域文化的怀旧，来追寻一种舒适的生活状态和精神状态，既体现了本土传统文化，又展现了现代时尚生活。

New World of Chinese Flavor

Analysis of "Xintiandi Development Mode" of Shui On Land

在世界，很中国

——解读瑞安"新天地模式"

Standing in the Shanghai Xintiandi of numerous luxury goods stores, fashion boutiques, foods and entertainments, you may wonder if you're still in China, because there are so many people of different skin colors from all over the world, and no matter what the season and weather is, the streets are always filled with crowds from day to night, full of Shanghai charm. Meanwhile, this place is also full of strong Chinese flavors, such as the historic relics of the First Congress of CCP among the old foreign-style houses, costing RMB 300 million for restoration. Passing the new and old buildings which witness the past, present and even future of the city, people can shop, drink, date, or even work here. "Yuppie lifestyle" is the high-end living concept deliberately created by it.

In 1997, a financial crisis broke out in Asia. When many foreign companies withdrawn their funds from Shanghai, the Shui On Land from Hongkong stayed and, with creativity and painstaking effort, made a great success in Chinese commercial property industry.

When many people still think "Shanghai Xintiandi" as a successful case of opportunism, the successive launch of "Xintiandi" series in Hangzhou, Chongqing. Wuhan, Dalian, Foshan and many other cities, proves that the group has already developed a reproducible development mode.

Location — Regional Core Cities or First-tier Cities

All programs are located at first-tier cities or regional core cities, such as Chongqing, Wuhan, Shanghai, and so on. Those cities own profound history and unique regional culture. The sites are always in city core area or old downtown area. There are representative historical or cultural relics on the site. The newly developed programs are featured with great scale and highly cost.

Planning — Using Retail Business to Open the Market and Build the Brand Awareness

The development sequence of general market is properly organized as: retail; residence; office building; residence phase 2 program; hotel & residence phase 3 program; retail; office building phase 2 program & international school; residence & office building & retail.

The planning creates urban complexes which include retail, luxury hotel, office building and high-quality residence. It retains valuable buildings and plants of the lands, and creates core value for the projects by restoring the old buildings. There are generally 30,000-50,000 m^2 malls.

Operation — Using Unified Leasing to Separate the Managers from the Operators

The operation insists on owning the property and using unified leasing to separate the managers from the operators. The objects are famous brands from all over the world, for instance, 85% of the 98 leaseholders in Shanghai Xintiandi are from regions beyond China. The high brand value not only greatly reduces the marketing cost for residence promotion, but gives the "Casa Lakeville" a good price much higher than surrounding residential programs.

Promotion — On-going Publicity

The brand makes use of modern means to strengthen the promotion of self image and improve its instant communication with outside world. A series of public activities are actively launched, organized and hosted, together with the help of media and participants, to expand the influence of the brand. Warm, healthy, active, stylish and dynamic image is established to improve the brand's affinity and cohesion. Constant publicity activities are made on core brands to bring more value, so as to improve the image and value of the surrounding areas.

Future — Green Products

The development concept of "Xintiandi" series is: "old buildings carrying forward the city spirit; new planning guiding the life style". Unlike other developers who are interested in undertaking "disruptive" developments in the city, Shui On Land has a special emotion for the "old buildings", and it reflects a unique humanistic concern. Every old building in "Xintiandi" is now blooming again.

The idea of "green architecture, green life" is infused into each "Xintiandi" program. In "Xihu Tiandi Phase 2 Project", sustainable design is fully reflected in construction and site location which pay much attention to sustainable factors, such as the improved water usages efficiency, optimization of materials and resources, decrease of energy cost and controlling of air pollution, and improved interior air quality.

In "Wuhan Tiandi", sustainable design such as green roof, double-layer hollow Low-E glass, water-saving installations, rain-water saver, carbon dioxide detector and energy-saving lightings are vastly used. To promote environment protection concept, an "E Exhibition Hall" is specially designed. As the first sustainable development education exhibition hall in Wuhan City, the "E Exhibition Hall" displays many sustainable designs and related information. With so many visitors attracted, it adds lively breath to the fashionable international metropolis.

Related Projects

Xihu Tiandi

Xihu Tiandi is very similar to Shanghai Xintiandi. So far, it is the only sister program by the group beyond Shanghai.

Ningbo Xintiandi

Ningbo Old Bund is located at the confluence of Yong River, Fenghua River and Yuyao River. By learning experience from Shanghai Xintiandi, the old bund retains local cultural buildings and builds many new buildings such as gallery, café, and museums. Thus it becomes the Ningbo Xintiandi.

Nanjing Xintiandi

Nanjing Xintiandi is planned to be built behind the "Nanjing Presidential Palace", next to the cultural street on Changjiang Road. There are 17 small buildings left from the Republic of China.

Suzhou Xintiandi

The Xuanmiaoguan Square located at the south gate of Guanqian Street will become the home of Suzhou Xintiandi. The land behind the city main road is about 24, 000 square meters in size, and the expected cost is about 250 million RMB.

Xiamen Xintiandi

Xiamen Xintiandi is expected to be located at the Lujiang Avenue. The project will start from the giant stone at the intersection of Shuixian Road and Lujiang Avenue, along Lujiang Avenue about 130 m to the south, and overlook the Gulangyu Island.

Fuzhou Xintiandi

"Rongqiao Xintiandi" is praised by Fujian media as a new version of "Xintiandi" in Fuzhou City. Located beside the Minjiang River, the project mainly covers a large themed park, leisure squares, sunken commercial streets and high-end multi-functional clubs.

Chongqing Xintiandi

Chongqing Xintiandi located on Nanbin Road is a waterside commercial street known as the "No.1 avenue" of the city. There are many historical relics reserved along the road.

在高端奢侈品商店林立、衣食住行和时尚娱乐一应俱全的“上海新天地”，你可能不会认为自己是在中国。来自世界各地的各色皮肤的人群穿梭于此，不管怎样的季节和天气，这里的白天总是人流如织，夜晚总是衣香鬓影，颇有大上海风韵。同时，这里也“很中国”——“一大”会址就在几栋斥资3亿元“修旧如旧”的老式洋房中间。从过去，到现在和未来，穿越重重新老建筑，人们可以在这里购物、小酌、约会、居住甚至办公。“走路可到达的雅皮士生活方式”，是“上海新天地”刻意营造的高端居住理念。

1997年亚洲金融风暴，当外资纷纷撤离上海时，来自香港的瑞安经过一番别有新意的打造和苦心经营，成就了中国商业地产的创新典范。

当很多人认为瑞安成功打造的“上海新天地”是一种机会主义的成功时，杭州、重庆、武汉、大连、佛山等地陆续成功开启的“新天地”系列项目，表明瑞安已经发展出一个可以有效复制的“新天地”模式。

选址——区域核心城市或一线城市

项目选址都是一线城市或区域核心城市，如重庆、武汉、上海、佛山、大连、杭州等。城市要有厚重的历史及独特的区域文化；地块选取位置都在城市的核心区域、老市区。地块内有城市较有代表性的历史文化建筑；新开发项目地块规模比较大，如武汉天地占地61万平方米，重庆新天地占地128万平方米；开发成本高，如“上海新天地”商业街占地面积不足3万平方米，投资却高达14亿元。

规划——零售先行，打造知名度

综合市场合理安排开发时序：零售——住宅——办公楼——住宅二期——酒店+住宅三期——零售——办公楼二期+国际学校——住宅+办公楼+零售。

打造城市综合体，融零售商业、高级酒店、写字楼、高档住宅于一体；保留地块有价值的建筑、植物；打造项目的核心价值，如上海的石库门，对旧建筑进行“造旧如旧”，但实现现代商业的运营功能；商业的产品模式一般为商业街、30 000～50 000平方米的商场。

运营——统一招租，实现管理者与经营者分离

只租不售，坚持持有物业，统一招租，实现管理者与经营者分离；招租对象均来自世界各地的知名品牌，如在“上海新天地”现有的98家租户中，85%来自中国以外的国家和地区。正是由于极高的品牌价值，不仅大大降低了在推广住宅时的营销成本，而且使“翠湖天地”一举成为上海顶级豪宅，售价遥遥领先于周边的项目。

推广——持续性的整体公关

利用现代手段加强自身形象的推广以及与外界的即时沟通。积极发起、参与、组织、承办一系列各种规模的公众活动，借助媒体与参加者的宣传，扩大“新天地”的影响；树立热情、健康、活跃、时尚、有活力的形象，增加亲和力和凝聚力；通过对核心品牌的不断宣传使自身品牌的含金量不断增值，带动周边形象与价值的提升。

未来——绿色产品

“以老建筑延续城市精神，以新规划引导生活方式”是“新天地”系列的开发理念。相比许多开发商乐于对城市进行“颠覆”式开发，“新天地”的“老建筑”情结有着与众不同的人文关怀。“上海新天地”里的石库门、“西湖天地”里的老建筑、“武汉天地”里的老牌坊、“重庆新天地”里的吊脚楼、“佛山岭南天地”里的祖庙，都在重新焕发青春。

在“新天地”系列的每一个作品中，都注入了“绿色建筑、绿色生活”的理念。如在“西湖天地”二期的建设中，充分体现了可持续设计系统：项目选址时注重可持续性因素，改善与提高用水效率，优化材料与资源的利用，减少能耗，控制大气污染，改善室内空气质量。

在“武汉天地”中，大量采用了屋顶绿化、双层中空低辐射玻璃、节水装置、雨水回收、二氧化碳探测器、照明节能灯具等。为了宣传环保理念，特别在项目中设立了“E展厅”。“E展厅”是武汉第一个可持续发展教育展厅，展示了废旧电池回收、自行车发电机、雨水处理示意图等，吸引了不少人前来参观学习，为国际化时尚都市增添了勃勃生机。

相关链接

西湖天地

西湖天地是上海新天地“最正宗”的姊妹篇，也是目前瑞安集团在上海之外做的唯一一个类似项目。

宁波新天地

宁波老外滩位于甬江、奉化江和余姚江交汇处，是1842年《南京条约》开辟的五个通商口岸之一，比上海外滩开埠早20年。宁波老外滩借鉴了上海新天地商业区的经验，依照原状保留了多处文化建筑，兴建了画廊、咖啡吧、展馆等一批新建筑。

南京新天地

计划中的新天地选址在旅游景点“总统府”背后，毗邻南京市正在重点打造的长江路文化街，拥有17幢民国时期的小楼。

苏州新天地

位于苏州观前街南门的玄妙观广场，将成为苏州的“新天地”。这是一块位于城市主干道约2.4万平方米的区域，预计投资额约2.5亿元。

厦门新天地

厦门思明区计划将厦门鹭江道商业景观配套工程打造成厦门的“新天地”。该项目从水仙路路口与鹭江道交叉口的大石头所在地开始，沿鹭江道向南约130米，隔鹭江道与鼓浪屿遥遥相对。

福州新天地

福州“融侨新天地”被福建媒体誉为“新天地”在福州的另一版本。“融侨新天地”商业会所位于闽江沿岸，主要由2万多平方米的大型立体式主题公园、休闲广场、300米下沉式商业步行街和高档多功能会所组成。

重庆新天地

有重庆新天地之称的“第一大道”是重庆南滨路上一个富有江浙水乡韵味的滨江商业街区。沿线保留有十九世纪四五十年代法国、美国、英国的水兵营以及众多的外国领事馆旧址，还有相对完善的基督教堂，具有一定的历史文化积淀。

Lingnan Tiandi, Foshan
佛山岭南天地

街区背景与定位
Street Background & Market Positioning

History 历史承袭

Lingnan Tiandi in Foshan is the first project that Shui On Group has developed in Guangdong. It assembles commercial street, high-end villas, foreign-style houses, apartments, office buildings, market and star hotel within 65 hectares (net site area is 52 hectares). Its gross building area is 1,500,000 m^2 .The complete time of its integrated development is expected to be 2020.

Foshan is a national historic city with culture connotation of five thousand years and over one thousand years' history since the city was founded. Ceramic, foundry, textile and Chinese patent drug used to be the mainstay industries of Foshan. Besides, Foshan is the birthland of Cantonese opera and a significant place of origin of Cantonese culture. Therefore, it occupies profound cultural connotation.

佛山岭南天地是瑞安房地产集团在广东省省内首个开发的项目，集商业街区、高端别墅、洋房、公寓、写字楼、商场、星级酒店等于一体，占地面积达65公顷（净用地面积52公顷），总建筑面积达150万平方米，预计整体开发完成时间为2020年。

佛山是国家历史文化名城，有着5 000多年的文化积淀，“肇迹于晋，得名于唐”，建城有1 300多年的历史。明清时期，陶瓷、铸造、纺织、中成药曾是佛山的支柱产业。佛山工商业、手工业繁荣兴盛，是全国“四大名镇”之一和“天下四聚”之一。佛山又是粤剧的发源地、武术之乡和民间艺术之乡，是“广府文化”的重要发源地，具有深厚的历史文化底蕴。

Location 区位特征

Lingnan Tiandi is located at Zumiao Donghuali area that is a traditional commercial center as well as a core area of historic culture with the densest cultural relics and historic sites of largest scale and most completed traditional scene. This area has witnessed the prosperous history of Foshan and preserved the essence of Lingnan culture. There are continuous valuable historic buildings, ancient gardens, ancient streets, old workshops, old shops, historic temple and former residences of celebrities and among them there are 22 cultural relics and historical sites.

岭南天地地处祖庙东华里片区，是传统的商业中心，也是佛山历史文化核心区，是文物古迹最密集、规模最大、传统风貌保存最完整的历史文化街区。这个片区见证了佛山辉煌的历史，保存了岭南文化的精粹，浓缩了昔日深厚的民俗文化、繁华的商业气息和鲜明的生活状态。这里有连片的颇具价值的古建筑、古园林、古街道、古作坊、古店铺、古祠庙及名人故宅，蕴含着极其丰富的历史文化资源，共有22处文物古迹，其中祖庙、东华里古建筑群是国家重点保护的历史建筑，也有其他广为人知的著名保护建筑，如简氏别墅、嫁娶屋、李众胜堂祖铺等。

Market Positioning 市场定位

Modern technique is applied in the development of Lingnan Tiandi to reform precious historic buildings of typical Lingnan residence style in Zumiao Donghuali area. Historical and modern elements are harmoniously integrated by infilling fashionable elements on the basis of keeping original style and feeling of historical buildings. Thus, the positioning of first phase of Lingnan Tiandi is to make the Lingnan Tiandi a commercial street that assembles international social intercourse, fashionable entertainment, boutique food and culture.

岭南天地在开发中运用现代手法，改造祖庙东华里片区具有典型岭南民居风格的珍贵历史建筑。在保留历史建筑原有风情的基础上，注入现代时尚元素，让历史与现代元素和谐地融合。因而岭南天地一期定位是以文化、旅游、商业和休闲为支柱，将岭南天地打造成为集国际社交、时尚玩乐、精品美食、文化荟萃为一体的商业街区。

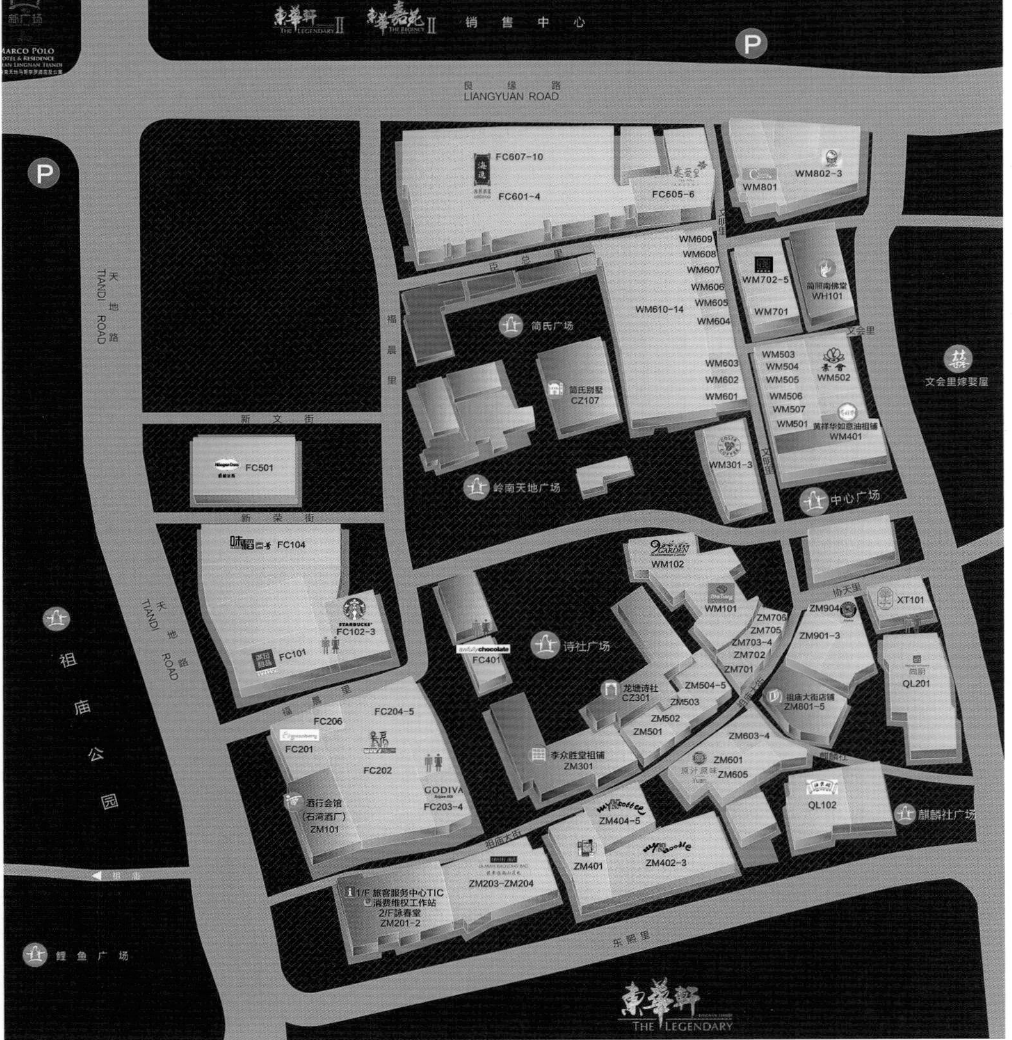

美食佳肴 Dining & Beverage

店号 Shop No.	商店 Shop Name
FC101	满记甜品 HONEYMOON DESSERT
FC102-3	星巴克 STARBUCKS
FC104	味稻一号
FC201	greenberry
FC202 ,FC204-5	吴系茶餐厅 Wuu's Hong Kong Cuisine
FC401	awfully chocolate
FC501	哈根达斯 Häagen-Dazs
FC601-4, FC607-10	海逸酒家 Harbour Plaza
FC605-6	泰爱里 Thai Alley
QL201	尚厨
WM101	芽庄
WM102	九号花园 NO.9 GARDEN
WM301-3	COSTA COFFEE
WM502	素会
WM801	Chéris Patisserie
WM802-3	Hello Kitty Cafe
XT101	大树吧 Big Tree Pub
ZM203-4	翡翠拉面小笼包 Crystal Jade La Mian Xiao Long Bao
ZM401	糖仁街 Chinatown Dessert
ZM402-3	my noodle
ZM404-5	my coffee
ZM601,ZM605	原汁原味
ZM904	柏龙PAULANER

生活精品，配饰及其他 Lifestyle, Accessories and Others

店号 Shop No.	商店 Shop Name
FC206	沁花园 PROVENCE
QL102	海棠阁
WM501	Lovely Lace
WM503-5	UNEEI
WM506	法蓝瓷 FRANZ
WM507	琉园 tittot
WM601	LOTOPIA
WM602	碧玉丰 BI YU FENG JEWELRY
WM603	富山香堂 FushanKodo
WM604	谭木匠
WM605	HARMAN
WM606	泰裕玉器
WM607	山川
WM608	玩味
WM609	岩陶
WM701	石湾美陶
WM702-5	悦画廊
ZM201-2(2F)	岭南天地詠春堂 WingChun MAA
ZM501	封伟民艺术工作室
ZM502	南粤广绣
ZM702	苏茛记
ZM705-6	天长地久 TITUS
ZM901-3	雅致LLADRÓ

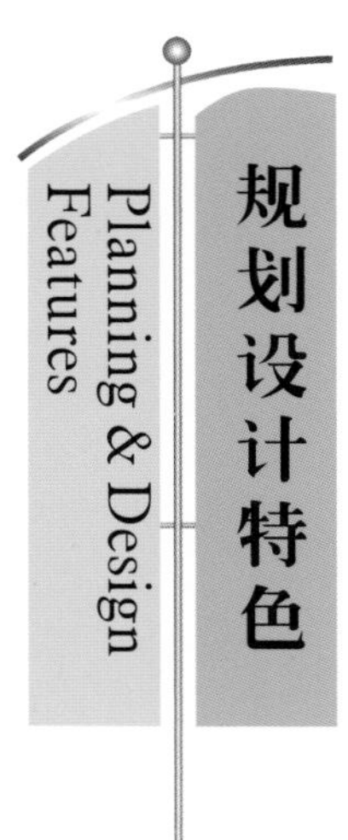

规划设计特色 Planning & Design Features

Street Planning 街区规划

The general planning of Lingnan Tiandi hammers at creating a new developing mode for Zumiao Donghuali area and reasonably integrating future development with the culture of this old city. The developing density of the planning design will be able to support this area to be a new, traffic-oriented and multifunctional city center. It is expected to transform the Zumiao Donghuali area to an attractive, lively city center that integrates work, life, shopping and entertainment through such a brand new mode of combining history protecting and city reviving.

Therefore, a "hill and valley" concept is put forward in the development of Lingnan Tiandi area. This concept means to keep historical buildings and preferable spatial scale in the center area of this project and conduct larger scale development of higher density at the edge zone of the project. All the development should respect the scale and historic feature, use part of the land for commercial, retail, entertainment, culture and art on the basis of keeping its original residential function, and eventually create an active modern area to release the developing stress of the old district. Moreover, according to the design concept of "hill and valley", high buildings are moved out from historical area gradually to avoid influence on sunlight and vision.

佛山岭南天地的总体规划致力于为佛山老城区的祖庙东华里片区创造一个新的开发模式，将未来的发展合理地融入具有悠久历史的城市文脉中。规划设计的开发密度将足以支持该片区成为一个新的、以公共交通为导向的、多功能的城市中心。期待通过这样一种全新的历史保护和城市复兴相结合的模式，将佛山的祖庙东华里片区转变为令人向往的，具有活力的，集工作、生活、购物和娱乐为一体的城市中心。

为此，在岭南天地的开发中提出了“山丘和山谷”的概念，在基地中心区保留历史建筑和较好的空间尺度，在基地边缘地带进行较高密度和较大尺度的开发。在历史建筑的保护上，保留祖庙和东华里民居建筑群。恢复祖庙的历史形象，控制建筑高度以保留其著名的屋顶形象和天际轮廓线。在尊重历史街区的尺度和特色的基础上进行新的填充和开发，激发历史街区的活力，在保留原本的居住功能的同时将一部分用地用于商业、零售、娱乐、文化、艺术和开放空间。最后，开发设计出一个具有活力的现代化区域，减轻历史老城区的发展压力。利用山谷形态的设计概念，将高楼逐步移出历史区域，避免其对日照和视线的影响。

Street Design Features 街区设计特色

Buildings in Lingnan Tiandi adopt typical Lingnan style which is primitive, exquisite and graceful. Maintaining original style, the design of the buildings also pays lots of attention to commercial elements to combine history and fashion harmoniously. The buildings use traditional materials and take reference of ancient pavilions to create fresh and bright image and highlight Lingnan character. Many unique craft features such as wooden sculpture, ceramics, lime sculpture and traditional doors have been adopted in the buildings. When visitors walk on the street, they can find no building or shop similar with others.

The façades of buildings are mainly built of grey bricks and black bricks collocated with multi-style glass and door frames. The roofs are multi-layer slope roofs made of traditional tiles of Lingnan. Parts of the eaves are decorated by colored paint and brick carvings. Exterior streets and stairs are built of old and new slab stones of different colors in different function area.

岭南天地中的建筑物基本沿用了典型的岭南建筑风格，整体格调古朴、玲珑、淡雅。在保留原有风情的基础上，又十分重视商业元素，历史与时尚在建筑风格中和谐融合。建筑也利用青砖瓦片、木梁石板等建造材料，创造通透空间及虚灵形体，形成清新明快的建筑形象，同时借鉴古代亭台楼阁原型，使新建筑千姿百态，凸显岭南气质。行走在街道中，基本上找不到相同的楼、店铺，建筑结构、店门变化万千，在建筑上采用了很多岭南独有的工艺特色，如木雕、陶瓷、灰塑、趟栊门等。

在建筑的外墙面上以深灰色、浅灰色为主，外墙主体以灰石砖、青砖堆砌而成，搭配多元化的玻璃和门框等元素。建筑屋顶常做多层斜坡顶，屋顶是岭南传统的瓦片，部分屋檐以彩画、砖雕等装饰，搭配和谐古朴。外部街道和阶梯主要以大小不一的石板地砖为主，用新旧混合的地砖堆砌成石板路，不同的功能区在材质和颜色上会有不同。

LOTOPIA
UNEEI

悦画廊

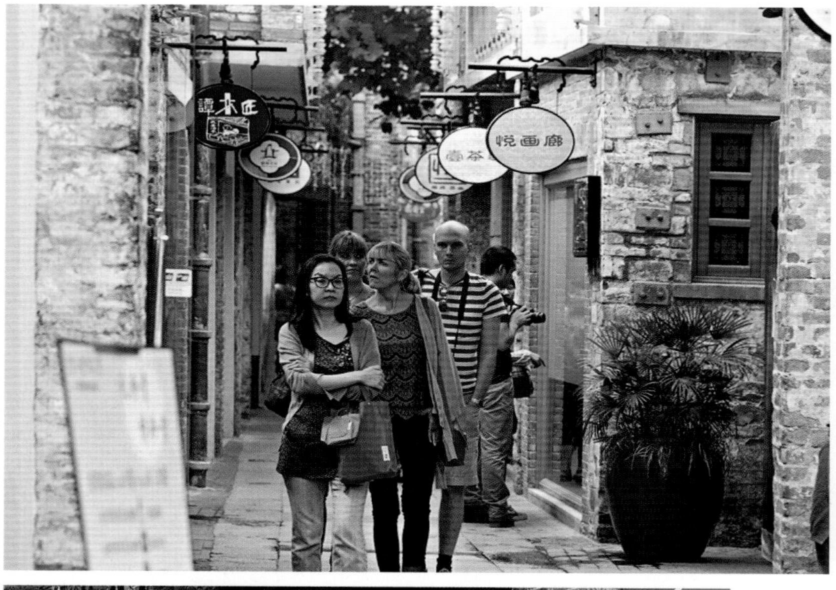

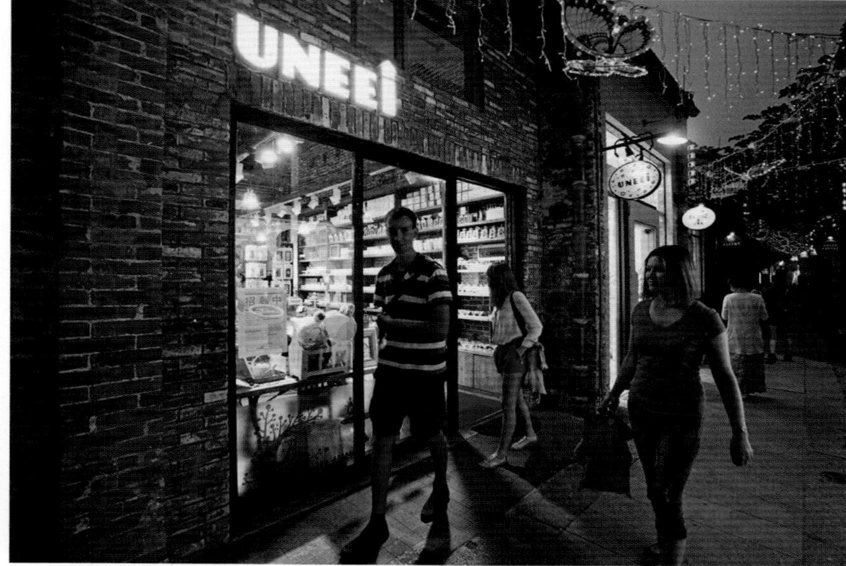

UNEEI

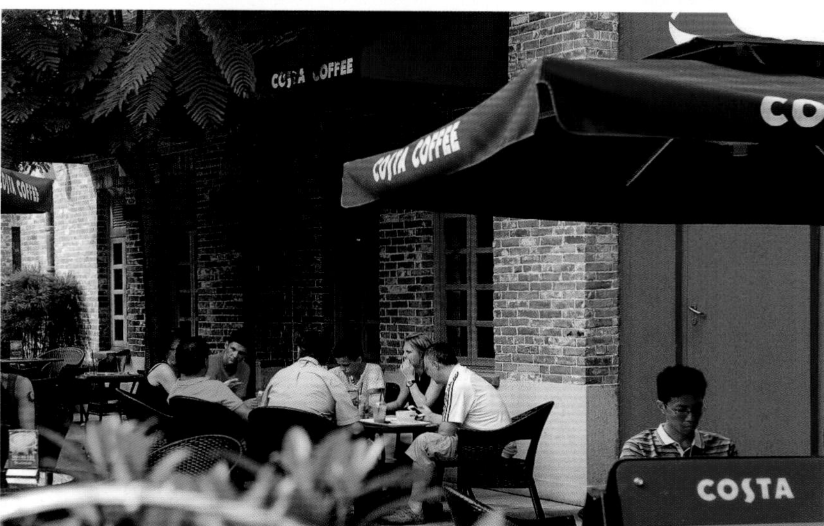

COSTA COFFEE
COSTA COFFEE
COSTA

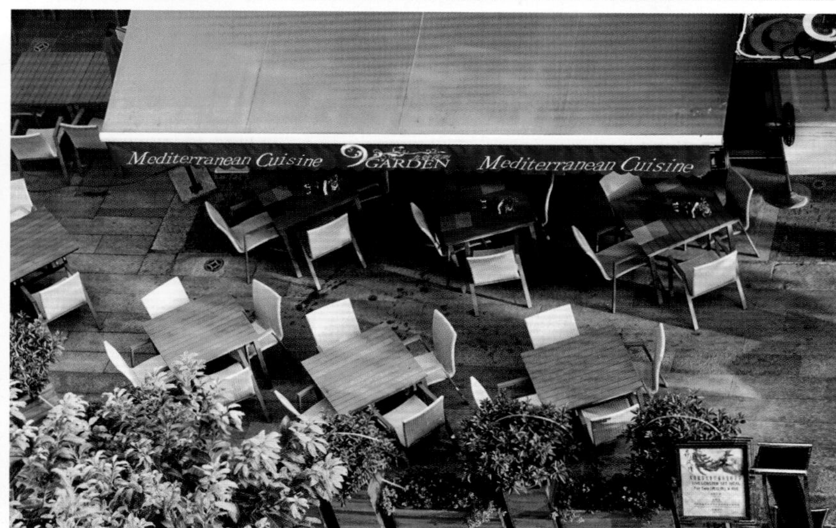

Mediterranean Cuisine
GARDEN
Mediterranean Cuisine

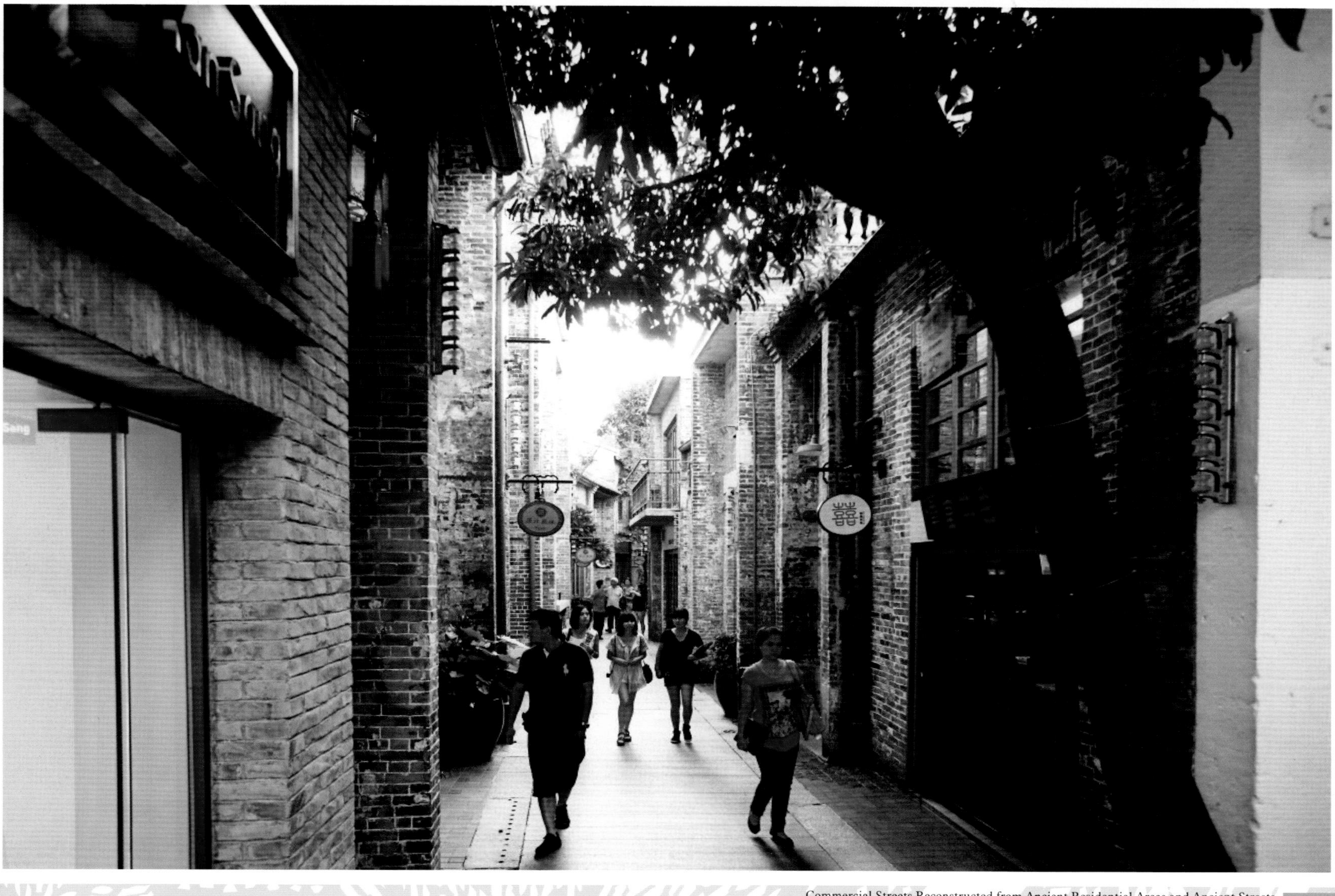

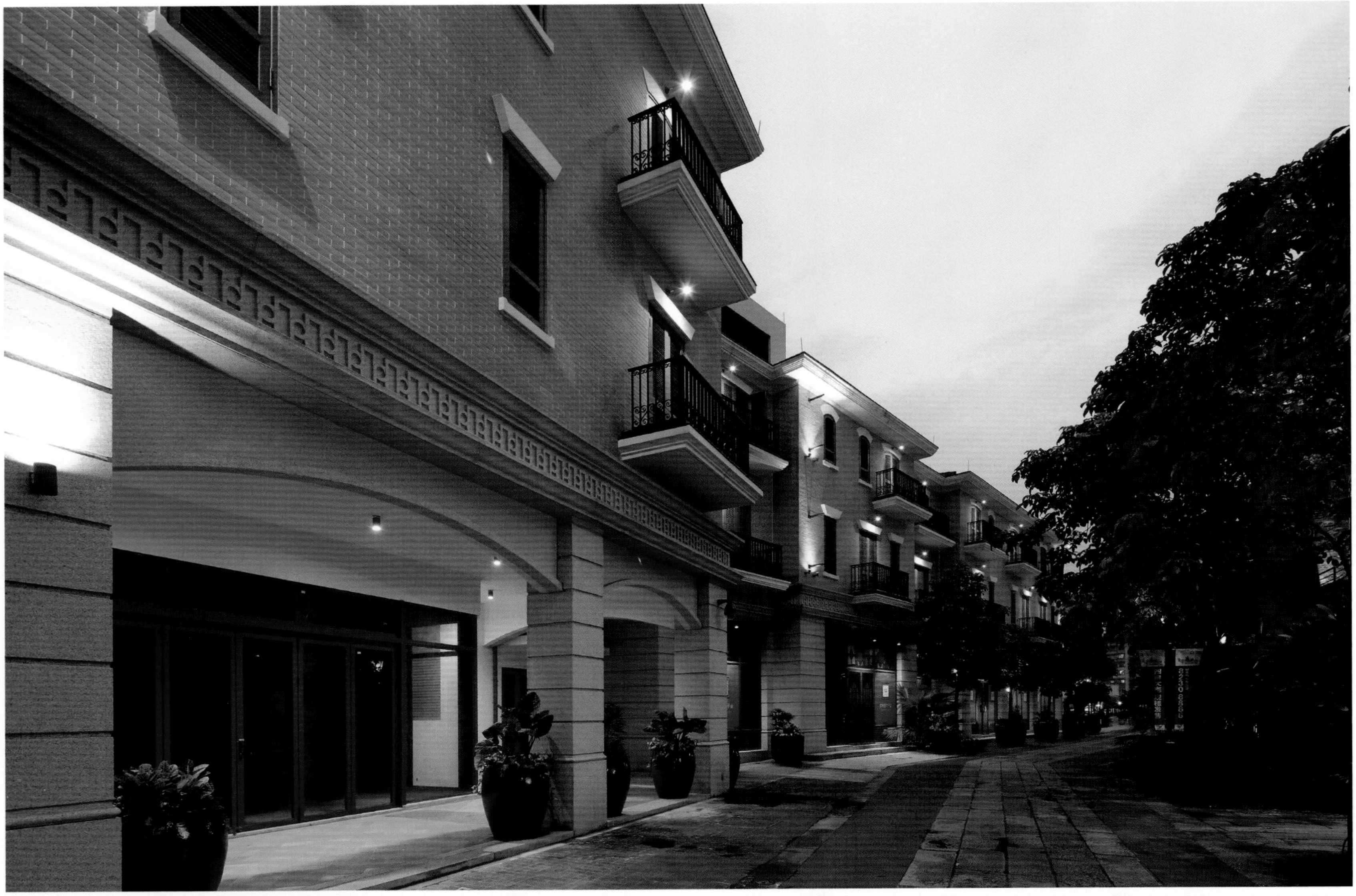

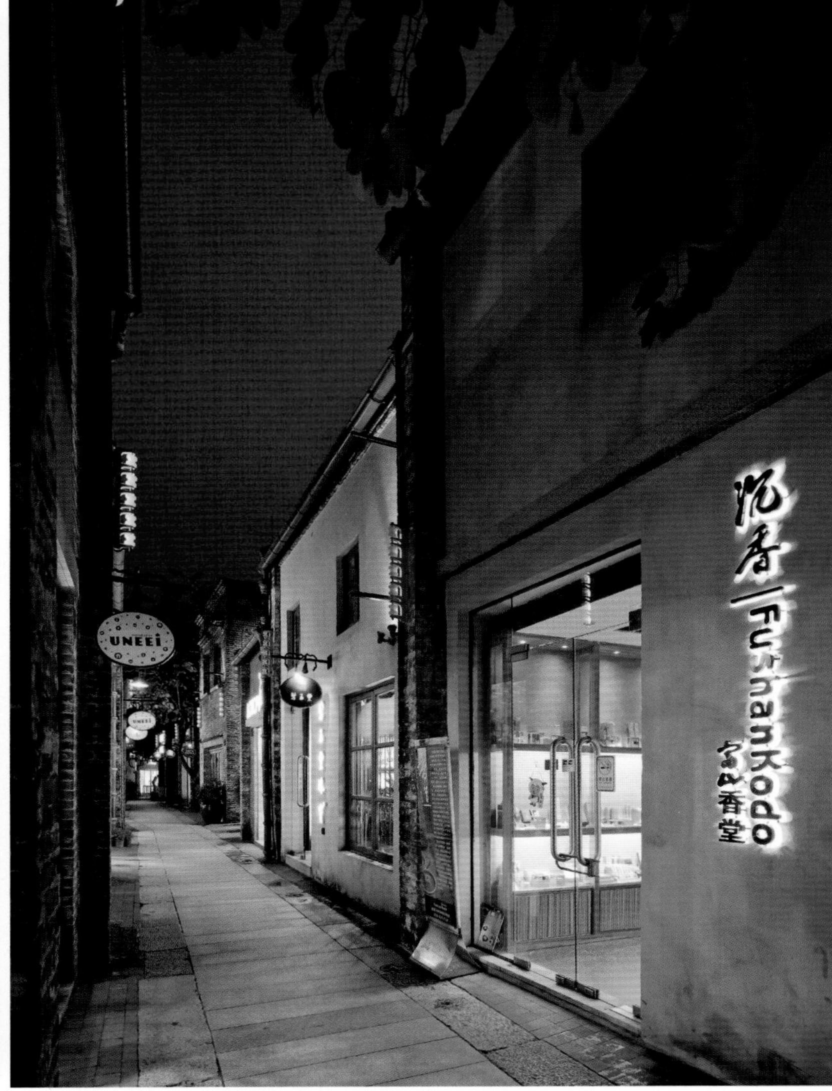
UNEEI
沉香
FushanKodo
富山香堂

Major Commercial Activities 主要商业业态

Commercial street of Xintiandi in Lingnan Tiandi is about 60,000 m². Its first phase area which is about 17,000 m² has fully opened. Most of the shops are international chains among which 70% are restaurants. Every building has its own story, and every shop has its own features. The Starbucks here is the only special decorated shop of Lingnan style in South China; the Costa here is the first Lingnan style shop of the world; and the largest Haagen-Dazs in South China is also full of Lingnan charm.

Besides, due to Foshan being a famous cultural town with ceramics, martial art, dancing lion and embroidery well-known around the world, the project has invited local artists to set up shops or galleries here. Now there are many art studios, shops and galleries.

岭南天地的新天地商业街共约6万平方米，一期商业街区现已全面开业，约17 000平方米，大部分为国际连锁商户，其中餐饮占7成，包括顶级粤菜馆味稻一号、9号花园地中海餐厅、泰爱里泰国餐厅、原汁原味中餐厅、海逸海鲜酒楼、翡翠拉面小笼包、吴系茶餐厅、柏龙德国餐吧、Chris法式甜品店及全国首家Hello Kitty主题咖啡馆等，即将开业的还有芽庄越南菜馆及商业街二期更多的特色餐厅。栋栋建筑有故事，间间店铺有特色，岭南天地星巴克是华南地区唯一一间以岭南风格装修的特装店，Costa咖啡店更是全球首间岭南特色特装店，而全华南面积最大的哈根达斯店也不乏岭南风味。

另外，鉴于佛山为有名的文化名镇，陶瓷、武术、醒狮、刺绣文化等蜚声海外，故项目更力邀本土著名艺术家设立门店或展示厅。游走于祖庙大街，可见佛山本土著名陶艺大师封为民工作室、梅文鼎大师的山川陶艺店、画家邹莉的悦画廊以及石湾美陶厂、海天酱油馆等，而叶问徒孙雷明辉师傅主理的咏春堂也已开业，为广大民众传授咏春拳法哲理及展示咏春文化。

Operation Measure 运营措施

Hand-over mode: The hand-over mode of the shops is "customized type", which means owner takes responsibility for planning the general layout and rebuilding the buildings based on the style of the project and then delivers the shops according to the different functions and demands. In this way, the project can keep the uniformity reasonably to the greatest extend.

Operation mode: In management, all the shops are only for rent, not for sale. All the decoration, operation and service are strictly managed and supervised by project company.

交付模式：商铺的交付模式为“定制式交付”，即由甲方统一规划整体布局。根据项目建设风格进行修葺、改建等，建筑物满足商业用途后，再根据每个商铺的功能业态和每个行业的不同需求，定制式交付品牌商家使用，最大限度并科学地控制规划的统一性。

经营模式：管理上，商铺只租不售，对商业街上租户的装潢、运营、服务项目，公司进行统一管理，严格监管。

Cakes & Desserts 糕点甜食类

Haagen-Dazs

Haagen-Dazs, which was founded in 1921 by Ruben Mattus, is a world popular ice cream brand. It uses pure natural materials without any antiseptics, artificial flavors, stabilizers and food pigments. It is tasty and healthy by skimmed milk and gains a good reputation.

The building of Haagen-Dazs is very unique in Lingnan Tiandi. The façade of the first floor is decorated with old-fashion colored brick collocated with yellow windows; big glass showcase is decorated with western white window frames and top hollowing carvings; and outside the door is the symbolic brand of Haagen-Dazs. The second floor is built of yellowwood and slope roof which produce a diversified feeling of Lingnan flavor. As for interior, in addition to pillars and walls keeping Lingnan style, all the others keep the Haagen-Dazs style.

哈根达斯

哈根达斯于1921年由鲁本•马特斯创建，是风靡世界的冰激凌品牌。它采用纯天然材料，不含任何防腐剂、人造香料、稳定剂和色素，用脱脂奶将美味与健康融合为一体，获得了"冰激凌中的劳斯莱斯"的美名。

佛山岭南天地中的哈根达斯店建筑风格别具特色，外墙部分一楼以旧式绘色砖块搭配黄木窗，大面积玻璃橱窗以西式白窗框加顶部镂空雕花作为装饰，门口处是哈根达斯的传统红色门框白色商标。一楼与二楼的连接处是灰白石质飘出楼台，挂着醒目的品牌商标。二楼则搭建黄木，配以顶部斜瓦，用多元化的形式搭建出岭南风情的建筑。室内装潢除了柱子与墙体保留岭南特色的灰色砖块方形柱外，其余装潢都保持哈根达斯店铺特有的天花吊灯、座椅风格和地面的材质拼接。

Honeymoon Dessert

Honeymoon Dessert is a well-known chain brand in Hong Kong which specializes in various kinds of traditional Hong Kong dessert, syrup and others. Founded in 1995, Honeymoon Dessert has gained praises from different customers after several years' improvement in management and innovation. Now, the branches of Honeymoon Dessert are spread all over the world.

Honeymoon Dessert at Lingnan Tiandi has a traditional Lingnan architecture appearance, while the interior decoration mainly submits to its own style. The walkway outside the shop is equipped with open-air seats.

满记甜品

满记甜品是香港地区知名连锁品牌，专门生产经营各类正统港式甜品、糖水及休闲类甜品。满记甜品集团创建于1995年，经过多年的独有家庭式管理和不断的研制创新，深得各方食客的认可，现在满记甜品分店遍布国内外。

满记甜品（岭南天地店）位于祖庙路岭南天地一号地块，外观上是传统的岭南建筑：两层，由旧式砖块堆砌而成，斜屋顶，小窗与门口等都沿用岭南风格，品牌标志挂牌与建筑相搭配。商铺外的走道上设有露天座位，增强临街的店铺标识，沿用满记甜品的深色系品牌风格。室内装潢完全以自身的品牌建设为主。

Chinatown Dessert

Hong Kong Chinatown Dessert was founded in 2009. The Chinatown Dessert in Lingnan Tiandi is the first flagship shop in Chinese Mainland. Its interior style is fashionable and simple. Chinatown Dessert pays great attention to food material quality and craftsmanship and tries to satisfy every customer. Now its featured products are Snow-ice, Magma Cake, Mango Tango and so on.

香港糖仁街

香港糖仁街创立于2009年，佛山岭南天地店是糖仁街在中国内地开设的第一家旗舰店，店内风格时尚简约、清新明快。糖仁街执着于食材品质及制作工艺，精心打造每一款甜品，致力于让每一位顾客在大快朵颐后都能获得满足。现在主打多口味的美味雪花冰、岩浆蛋糕“心太软”、芒果探戈等特色甜品。

Awfully Chocolate

Awfully Chocolate, featuring chocolate cake, is a cake shop founded in 1998 in Singapore. Its flagship shop in Lingnan Tiandi is a two-floor building with appearance of its usual black and white fashionable simple style.

奥芙莉

奥芙莉是1998年创于新加坡的一家蛋糕店，主打巧克力蛋糕。其佛山旗舰店位于岭南天地商业中心诗社广场，是一座二层的小楼房，门面装修是一贯的黑白时尚简约风格，二楼装修文艺清新，配上大露台，显得很有小资情调。

Food & Beverage 餐饮类

My Noodle

My Noodle is managed by Gao Wen'an. He has observed and studied the exterior decoration, food and service of the restaurant thoughtfully in detail. The façade of My Noodle inherits Lingnan architecture style. The outdoor seats and adornments, shape and colors of art wares, exquisite and unique droplights, all these are simple yet luxury.

My Noodle（高文安·面馆）

My Noodle由高文安先生经营，他将面馆从外部装潢到饮食与服务都做了详细而周到的研究。My Noodle的整体外部建筑保留岭南建筑风格，露天部分的座椅选材与各种摆设装饰都体现了高文安在设计中的艺术触觉。室内外工艺品的选用，形状与颜色的搭配运用，特别的雕塑，造型别致的吊灯，简单中透着低调的奢华。

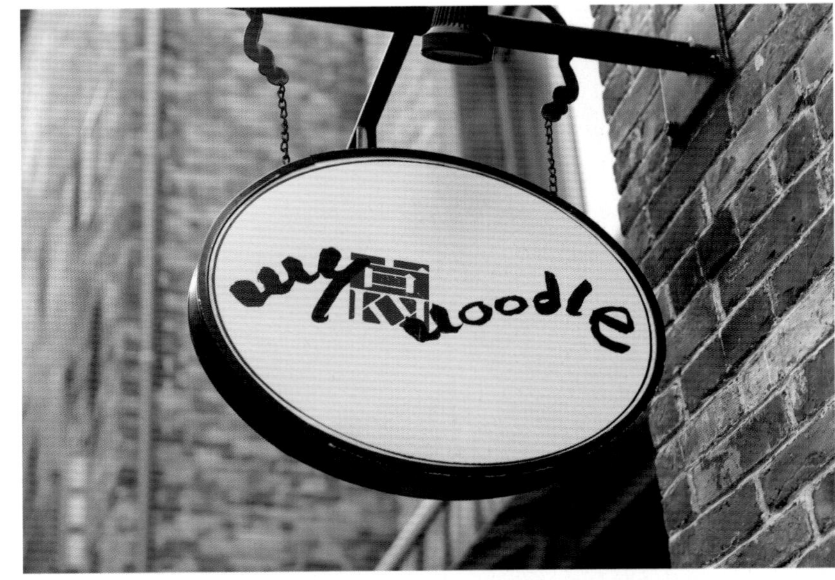

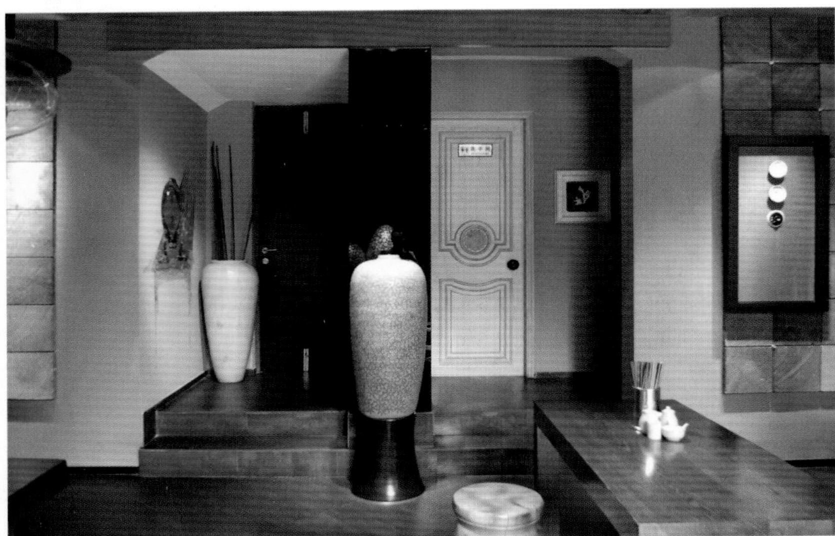

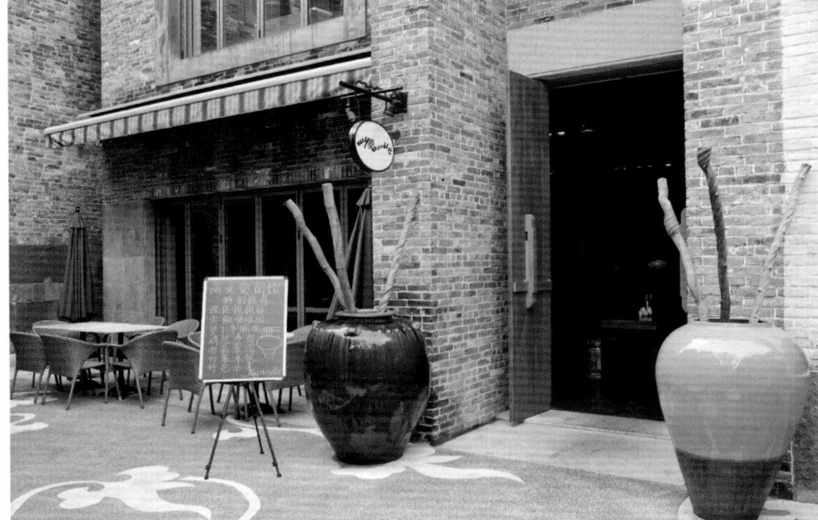

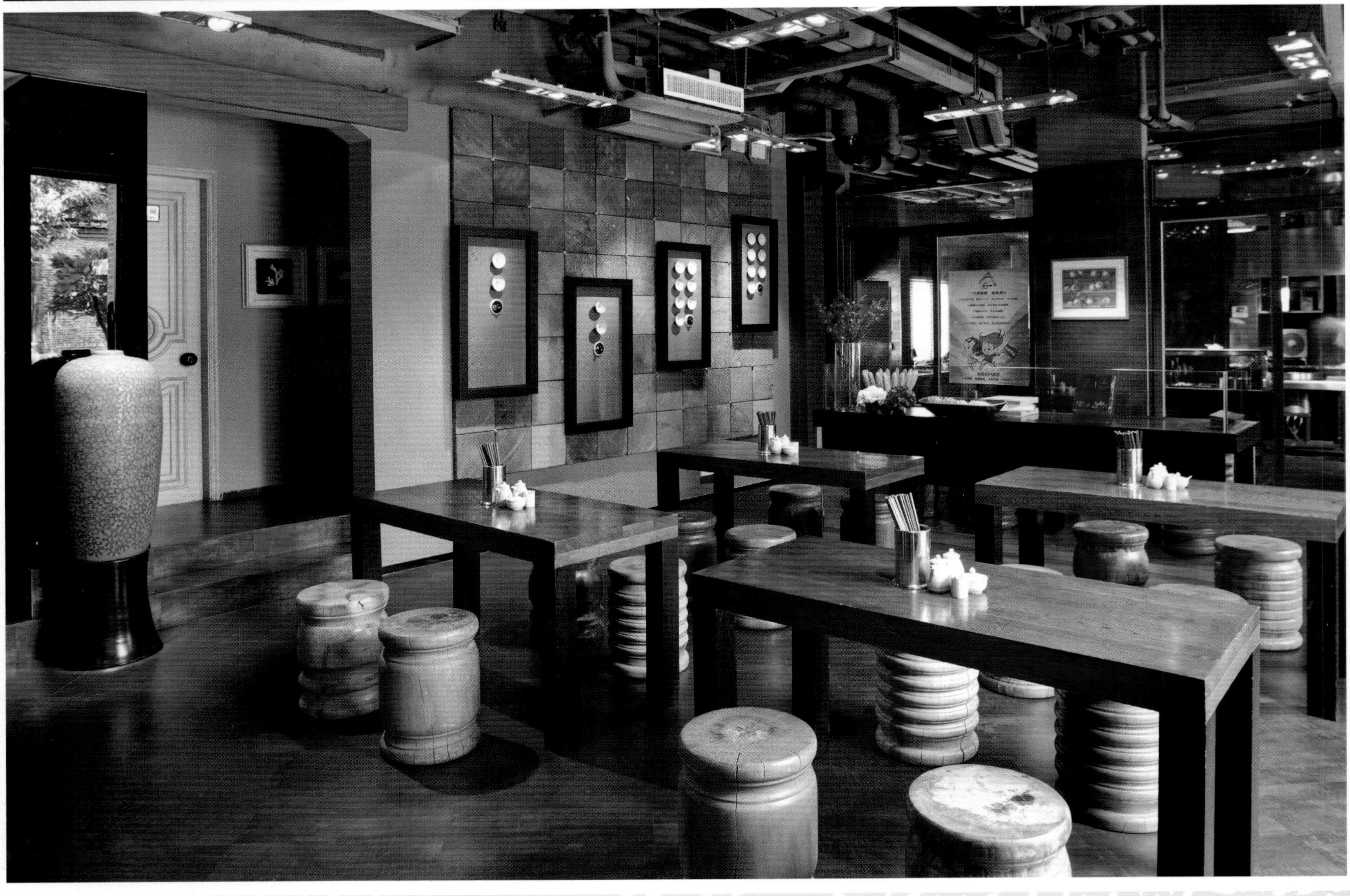

My Coffee

My Coffee is Gao Wen'an's favorite, the father of interior design of Hong Kong. Culture is the eternal theme here. Walls and adornments at corners of walls particularly agree with the style of the whole project. The interior design and furnishing create a comfortable family feeling and highlight a petty bourgeoisie taste.

My Coffee（高文安·咖啡馆）

My Coffee是“香港室内设计之父”高文安先生的心头爱，也是三大My系列（面馆、咖啡馆、健身馆）之一。文化是这里永远的主题。咖啡馆的墙面、墙角的装饰品特别符合项目的风格，尽显艺术气息。馆内设计及陈设本身所散发出的浓郁色彩吸引着人们的眼球，桌面上是书本摆设，营造出舒适的家的氛围，凸显出小资情调。

Pinweixuan • Shangchu

Pinweixuan · Shangchu is designed by Gao Wen'an. It provides Cantonese cuisine, Teochew cuisine and fashion Western-style food. It pays great attention to food quality, dining atmosphere and healthful environment to lead a healthy and fashionable diet trend. It innovatively integrates modern and traditional design elements to create a unique restaurant style which has a bright image and good taste.

品味轩 • 尚厨

品味轩•尚厨位于岭南天地商业中心东熙里。以全新理念装修落成的尚厨，注入了高文安先生的创作精髓，提供粤菜、潮州菜、时尚西餐等菜式，注重时尚餐饮质量的要求以及就餐气氛、环境洁净等元素，引领出一个健康时尚的饮食潮流。它融合了现代与传统的设计元素，对其组合创新，以一定的现代气息形成独特的餐厅风格，形象鲜明，品位非凡。

Starbucks

Starbucks in Lingnan Tiandi is the fifth branch of Starbucks in Foshan. The pure coffee culture of Starbucks perfectly integrates the history with this old city. The unique design of the shop brings customers a unique experience. The shop is an arcade building of Lingnan style which is decorated with traditional style and brings customers good mood when tasting coffee.

星巴克

星巴克（岭南天地店）是星巴克在佛山地区的第五家分店，位于祖庙商圈内，千年古城的历史积淀与星巴克纯正的咖啡文化完美交融，独具特色的门店设计为顾客带来了独特的星巴克体验和“第三生活空间”。它的门店是具有岭南风格的骑楼式外观建筑，雕花点缀着屋檐，楼身是青砖绿瓦，有着舒适宽敞的骑楼底，以龟背锦为花纹的窗台玻璃与大红灯笼相互辉映，为品尝咖啡的顾客带来愉悦的心情。

Medicine & Health 药店养生类

Eu Yan Sang Museum

The first Eu Yan Sang drugstore was founded by Yu Guang in Malaysia and was carried forward by Yu Dongxuan, the eldest son of Yu Guang. Now it is the largest traditional Chinese medicinal materials producer and retailer in Hong Kong, Singapore and Malaysia.

Located in Lingnan Tiandi, Eu Yan Sang Museum is the first retail museum concept store. There is a left door and a right door to get into the building. Entering from the left door, there is a gallery that exhibiting the historical and cultural relic of this old famous shop of hundred years. At the end of the gallery is an imitation of Eu Yan Sang drugstore of 1930s .

余仁生博物馆

首家余仁生药店由创始人余广在马来西亚创立，后来余广的长子余东旋把余仁生药店发扬光大。现在，余仁生已成为中国香港、新加坡以及马来西亚规模最大的传统中药材产品制造及零售商之一，业务遍布东南亚。

位于佛山岭南天地龙塘诗社的余仁生博物馆是余仁生药店的首家零售博物馆概念店，建筑有左右两个门，从左边门进去，迎面是古色古香的“时光隧道”——一条展示这家百年老字号百年历史照片和文物的廊道。穿过廊道，是一个仿照20世纪30年代风格设计的余仁生药铺：大红的灯笼、及臂的红木药柜、琵琶形盒子装着的秤……

Huang Xianghua Health Promotion

Huang Xianghua Health Promotion was founded during Xianfeng Period of Qing Dynasty by Huang Zhaoxiang, also named Huang Xianghua. Its "Huang Xianghua" medicinal oil is an old famous brand, by which Huang Zhaoxiang became a millionaire in Foshan. It set up branches around the nation and was later entrenched in Hong Kong. Now the shop, which is reconstructed according to its original style, is managed by descendents of Huang Xianghua. It has a traditional layout with outer room selling oil and inner room being health promotion room. Its interior decoration is graceful and antique.

黄祥华养生堂

黄祥华养生堂始创于清咸丰年间，创始人为黄兆祥（又名黄祥华），因经营药油成为佛山巨富，其所经营的“黄祥华”如意油是百年老字号，在全国各大城市开设分店，后扎根于香港。现在的店铺由其后人开设，祖铺现在以清代商业街“竹筒屋”式铺面重新迎客，整个店铺按照清代祖铺建筑风格进行了重新修缮，古香古色，原汁原味地反映了祖铺当年的风貌。店铺仍是传统的前铺后居式布局，外间卖药油，内间是养生堂。堂内布有褐色木质餐台和淡黄色木栅栏，清雅复古。

Flower Shop 花店

Provence

Provence is a very interesting flower shop with so many kinds of flowers, mini potted plants and dried flowers.

Provence

Provence是一家很有意思的鲜花店，里面有不少品种的花，如绣球花、厄瓜多尔玫瑰、郁金香……另外还有迷你小盆栽和干花。进入其间，众花争艳，真是“名花倾国两相欢，常得君王带笑看”。

Former Site of Longtang Poem Club 龙塘诗社旧址

Longtang Poem Club, which has over one hundred years' history, is a gathering place of scholars in Foshan during the late Qing Dynasty and the early Republic of China. After the Lingnan Tiandi project started up, it was repaired. It has a courtyard that surrounded by an ambulatory. There are two buildings lain as "L" shape. Both the main building and wing building are of exquisite traditional structure and decorations such as patterns of peony, thrush that mean wealth and luck. The northeast corner of the wing building is the entrance of the courtyard. The landscape wall in the north is decorated with lime sculpture and double-side colored painting.

龙塘诗社是清末民初佛山岭南地区文人活动的聚集地，至今已有百余年历史。佛山岭南天地项目启动后，对其进行了修缮，现在龙塘诗社院子的四个角耸立着四棵大树，龙塘诗社建筑及庭院景观墙与西面的李众胜堂祖铺相合，形成了一个由回廊建筑环抱的庭园，古树婆娑，环境雅致。单体建筑为房屋两栋，两栋建筑呈“L”形平面布局，主楼坐南向北，为两层硬山顶青砖木结构，走廊柱间为拱圈设计，雕饰精美灰塑图案，画着牡丹、喜鹊、画眉等象征富贵吉祥的花鸟；副楼构筑二层，坐东向西，主、副楼交接之间设雕花栏杆木楼梯。副楼东北角的门楼通向文明里外街，为庭院出入口，北面的景观墙上装饰灰塑和双面彩画。

The Jian's Villa 简氏别墅

The Jian's Villa is two-floor high with a gate tower, a main building, a back building, a theatrical stage, a storage building and a garden. The main building is an imitated Italian renaissance style mansion. Window glass, exterior walls of back building and tile carving on sunblind show a distinct Chinese style, while the general style is of Western style.

简氏别墅楼高两层，有门楼、主楼、后楼、戏楼、储物楼、花园等建筑，主楼是仿意大利文艺复兴时期府邸式建筑。别墅在窗玻璃、后楼外墙、窗檐砖雕上体现出明显的中式风格，而在楼房的主体风格、楼梯栏杆、西楼上又仿西洋建筑，体现出中西合璧的特色。别墅至今保存良好。

Lee Chung Sheng Ancestral Shop 李众胜堂祖铺

The shop is two-floor of brick and wood structure and in a traditional layout of shop downstairs and dwelling upstairs. The front of the shop is spacious. There was a garden at the east part which has disappeared now.

李众胜堂祖铺位于祖庙大街18号，是民国初年“保济丸”销路大开时所建的大型店铺。店铺为两层砖木结构，是下铺上居的传统布局。铺门前开阔，原有“保济滋大”的牌匾。铺东侧原设有花园和院落，现已不见。1998年，佛山市人民政府公布其为文物保护单位。

引导指示系统 Guidance & Sign System

Wuhan Tiandi
武汉天地

街区背景与定位
Street Background & Market Positioning

History 历史承袭

Wuhan Tiandi is elaborately built by Rui On Land Co., Ltd. It is located at Yongqing area of center district of Hankou. Its site area is about 610,000 m², and its total building area is about 1,500,000 m². Its total investment is about ten billion RMB. It is expected to be completely finished in 2014.

Yongqing Street is located at the south side of middle Huangpu Road and east side of Zhongshan Avenue, Jiang'an District. From the late Qing Dynasty to 1937, the Yongqing area is the main area of Japanese concession in Wuhan. During the Anti-Japanese War, Yongqing area became a center of Wuhan military and civilian to beat back Japanese invaders.

武汉天地是由瑞安房地产有限公司精心打造的项目，项目位于汉口市中心城区永清地块。项目占地61万平方米，总建筑面积约150万平方米（包括公共设施及公共空间），总投资额约100亿元人民币，预计2014年将整体落成。

永清街位于江岸区黄浦路中段南侧，中山大道东侧，街道长250米，宽16米，清末时，曾有四姓集资在此修“永清寺”。清末至1937年，永清地区为日本在汉口的主要租界区。抗日战争时，永清地区成为武汉军民抗击日本侵略者的中心。

Location 区位特征

Wuhan Tiandi is located at Yongqing area of Hankou, facing Changjiang River at the east and enjoying the picturesque scenery of Changjiang River. The project has a very convenient traffic, connecting with Jiefang Avenue, Yanjiang Avenue, Jinghan Avenue and Zhongshan Avenue and near to Huangpu Station of light rail.

武汉天地位于汉口城区永清地块，东临长江，面向风景如画的江滩公园，尽览长江美景。项目周边连接解放大道、沿江大道、京汉大道、中山大道，与轻轨黄浦站邻近，交通非常便利。

Market Positioning 市场定位

Wuhan Tiandi refers to the developing mode of reconstruction project of Shanghai Taipingqiao area and will be developed into a city center comprehensive project with multi-functional facilities of residence, office, hotel, retail, restaurant and entertainments to match with the whole long-term plan and development strategy of internationalizing of Wuhan.

武汉天地参照上海太平桥地区重建项目的发展模式，将打造成为集住宅、办公楼、酒店、零售、餐饮、娱乐等多种功能设施于一身的市中心综合发展项目，配合整个武汉国际化发展的远景规划和发展战略。

Street Planning 街区规划

Wuhan Tiandi is divided into Parcel A at the south of Huangpu Bridge and Parcel B at the north. Parcel A is mainly for comprehensive business office and parcel B is for dwelling. High-rise public buildings are arranged along the Jinghan Avenue. In the planning layout, there are 43,600 m^2 concentrate greening and three greening corridors perpendicular to the river.

Its general layout, considering the characters and function structure of the concession of old Hankou, adopts reconstruction methods as follow.

1. To continue the old city texture.

The site of this project is located at the concession of old Hankou. The planning layout fully considers features of the layout, scale and building heights and creates an amiable street space, therefore, the function of the old town is vitalized, and meanwhile the texture is continued.

2. To increase the density of roads to ease traffic stress of the area.

The client invites Atkins Consultant Co., Ltd. to do traffic assessment and adopts the density of the old street to create pleasing space and to ease the traffic stress.

3. To enrich landscape form in the city space.

The design does three-dimensional model analysis of important landscape points to give a series of guiding opinion on the height, width, color, roof design and so on.

4. To build its own environment, meanwhile provide a large area of concentrate greening space.

The project is facing Jiangtan Park. In order to strengthen the connection between the project site and Jiangtan Park, light rail station, three greening landscape corridors are planned to make full use of natural landscape resources and improve part of microclimate.

5. To create the cultural environment of the project by combining historic buildings.

Combining eight preserved historic buildings at the project land, a 25,000 m^2 historical featured commercial area - Wuhan Tiandi is built. The planning design fully explores the cultural feature of Wuhan and combines the historic buildings to improve the function of the area and create its cultural environment landscape.

6. To create a new landmark of Hankou and enrich the skyline.

Super high-rise building of about 297 meters high is arranged at the northwest corner of the land to create a new landmark of Hankou and lead the landscape form of this area.

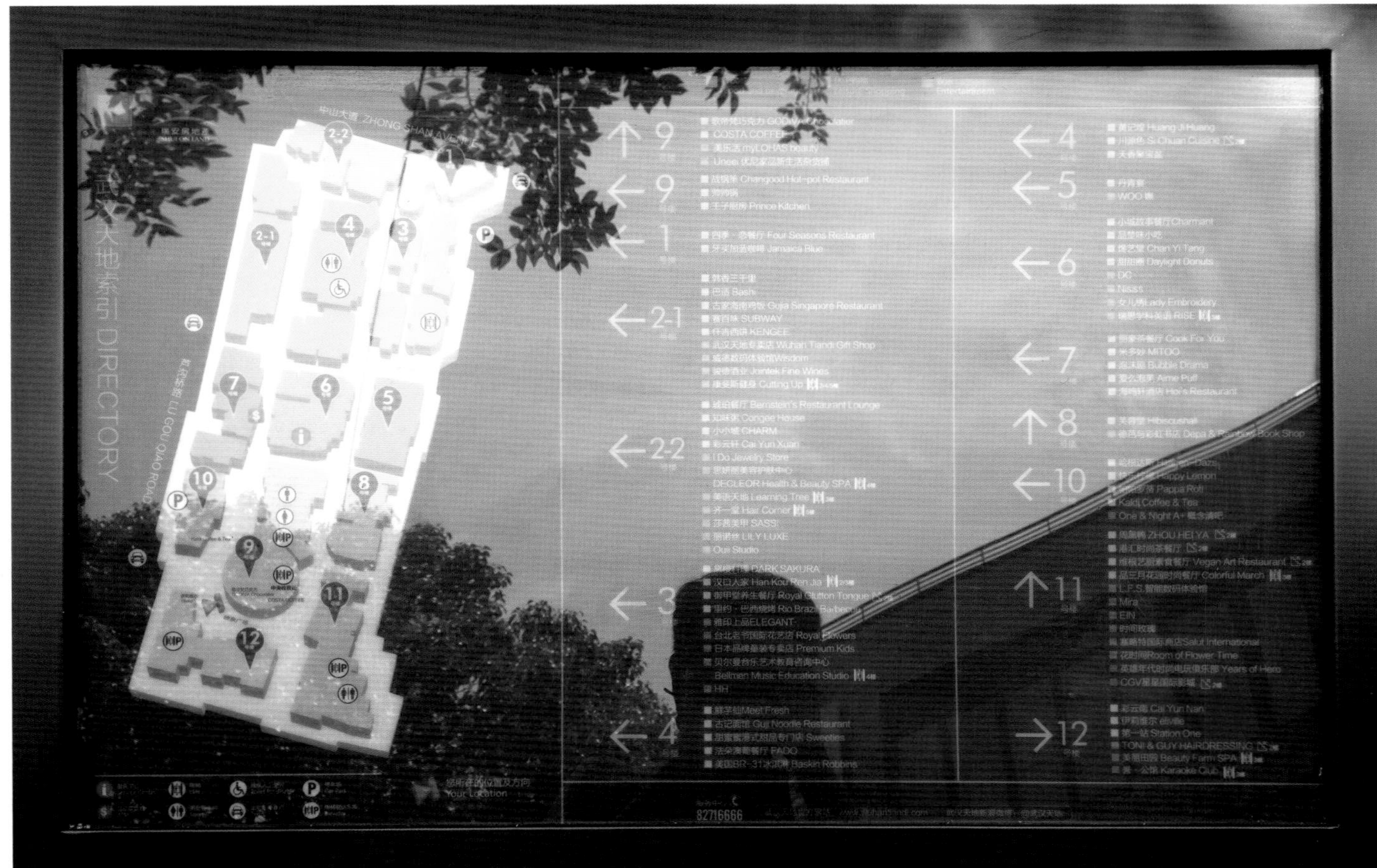

武汉天地项目以黄浦大桥为界，大桥以南为A地块，以北为B地块。A地块以综合商务办公为主，B地块以居住为主，高层公共建筑主要沿京汉大道布局，其中在用地西南角规划约300米高的超高层地标建筑。在规划布局中规划4.36公顷的集中绿化区和三条垂直于江面的绿化通廊。

在整体布局上，结合武汉老汉口租界区的布局特色与功能结构的分布情况，采用了以下几种改造方式。

1.延续旧城肌理

该项目用地位于老汉口的租界区内，规划布局充分考虑旧城的布局方式、街道尺度、建筑高度等方面的特色，力求在完善功能配置的前提下进行规划布局。该方案采用小街坊尺度、围合式布局，塑造亲人的街道空间，在盘活旧城功能的同时较好地延续了旧城肌理。

2.提升路网密度，缓解地区的交通压力

规划方案邀请阿特金斯顾问有限公司进行项目的交通评估，同时该方案采用老城街道密度，在塑造宜人空间尺度的同时缓解地区的交通压力。

3.丰富城市空间景观形态

规划方案设计选取重要的景观视点进行三维实景模拟分析，在建筑沿江天际线、上下二桥建筑空间形态、建筑高度、建筑体量、建筑后退道路红线、建筑色彩以及屋顶设计等方面提出一系列具有指导性的意见。

4.在塑造自身环境的同时提供大面积的集中绿化空间

项目用地临汉口江滩公园，为加强项目用地与江滩公园以及轻轨站点的联系，规划布局了垂直于江滩的三条绿化景观通廊，在充分利用自然景观资源的同时顺应武汉市的风向，改善了局部小气候环境。

5.结合历史建筑进行项目的人文环境塑造

结合项目用地内8栋保留的历史建筑，塑造2.5公顷的历史风貌特色商业区——武汉天地，规划设计充分挖掘武汉的人文特色，并结合历史建筑进行规划设计，在完善功能配置的情况下，塑造了该区域的人文环境景观。

6.塑造汉口新地标，丰富天际轮廓线

项目用地的西北角规划布局约297米的超高层建筑，结合汉口大门的规划理念，塑造汉口新地标，统领区域的景观形态。

Street Design Features 街区设计特色

Wuhan Tiandi is composed of twelve buildings including eight historical buildings and brick buildings not exceeding three floors, to create a spacious and comfortable commercial space. The facades of the buildings keep brick wall and roofing tile of old days, while their interiors are decorated according to modern life of urbanites. Buildings of this project are a mixture of preserved buildings, new buildings of old style and fashionable steel structure buildings with glass windows, supplemented with pavement, chairs, sculptures, plants, fountains, works of art and so on to reveal the leisure shopping atmosphere of the whole street.

武汉天地由12栋建筑组成，保留区域内8栋历史建筑，配以不超过3层的砖砌低矮建筑，营造出开阔、舒适的商业空间，建筑外部保留了当年的砖墙、屋瓦，但每栋建筑内部均按现代都市人的生活方式量身定做，体现出现代休闲的气氛。

项目的整体建筑风格由保留建筑、沿袭保留建筑风格的新建筑和时尚钢结构玻璃窗建筑三种风格穿插混搭而成，辅以街道铺装、休息座椅、小品雕塑、绿化、喷水池、艺术品等，体现出整个街区的休闲购物氛围。

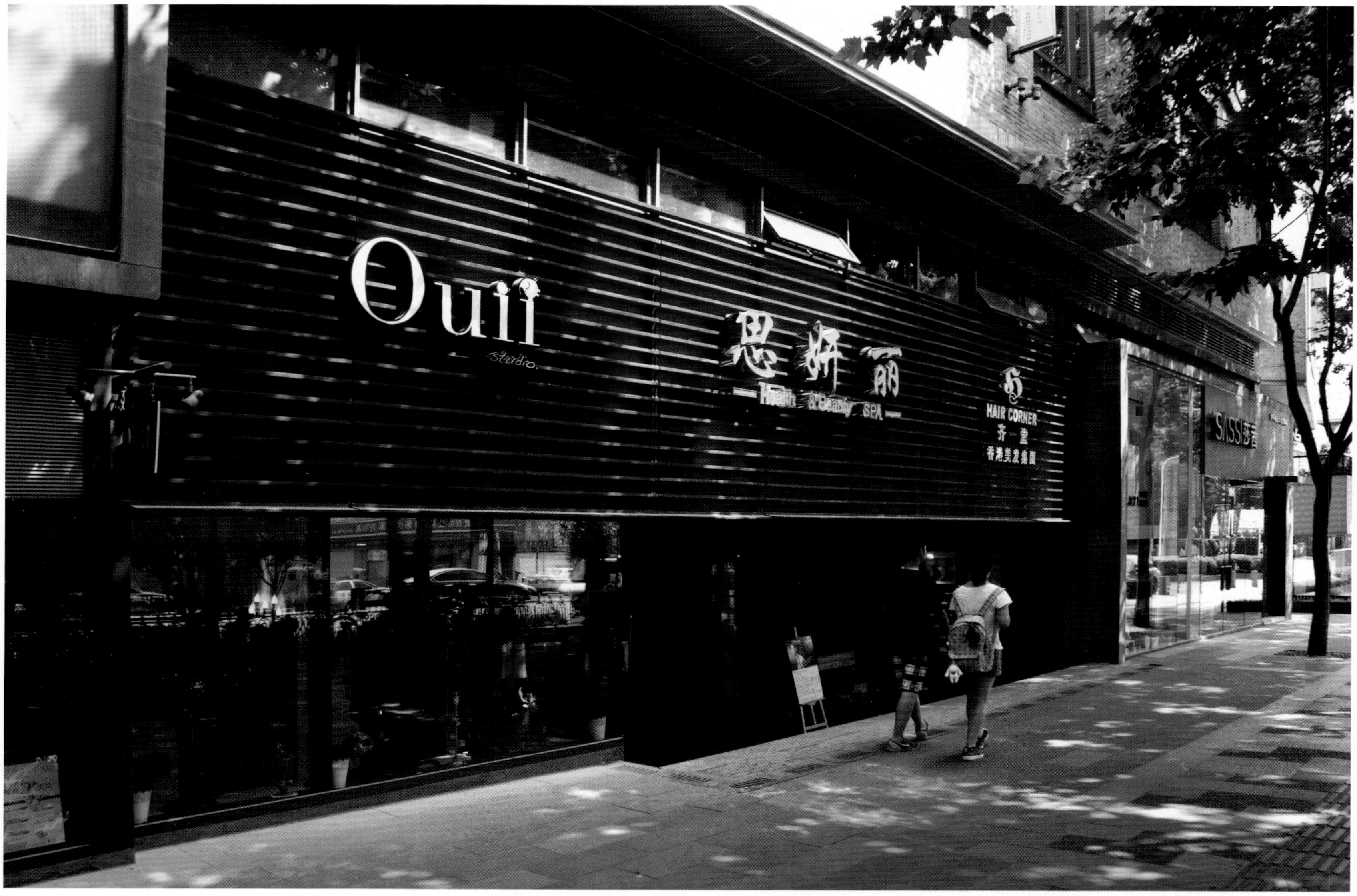

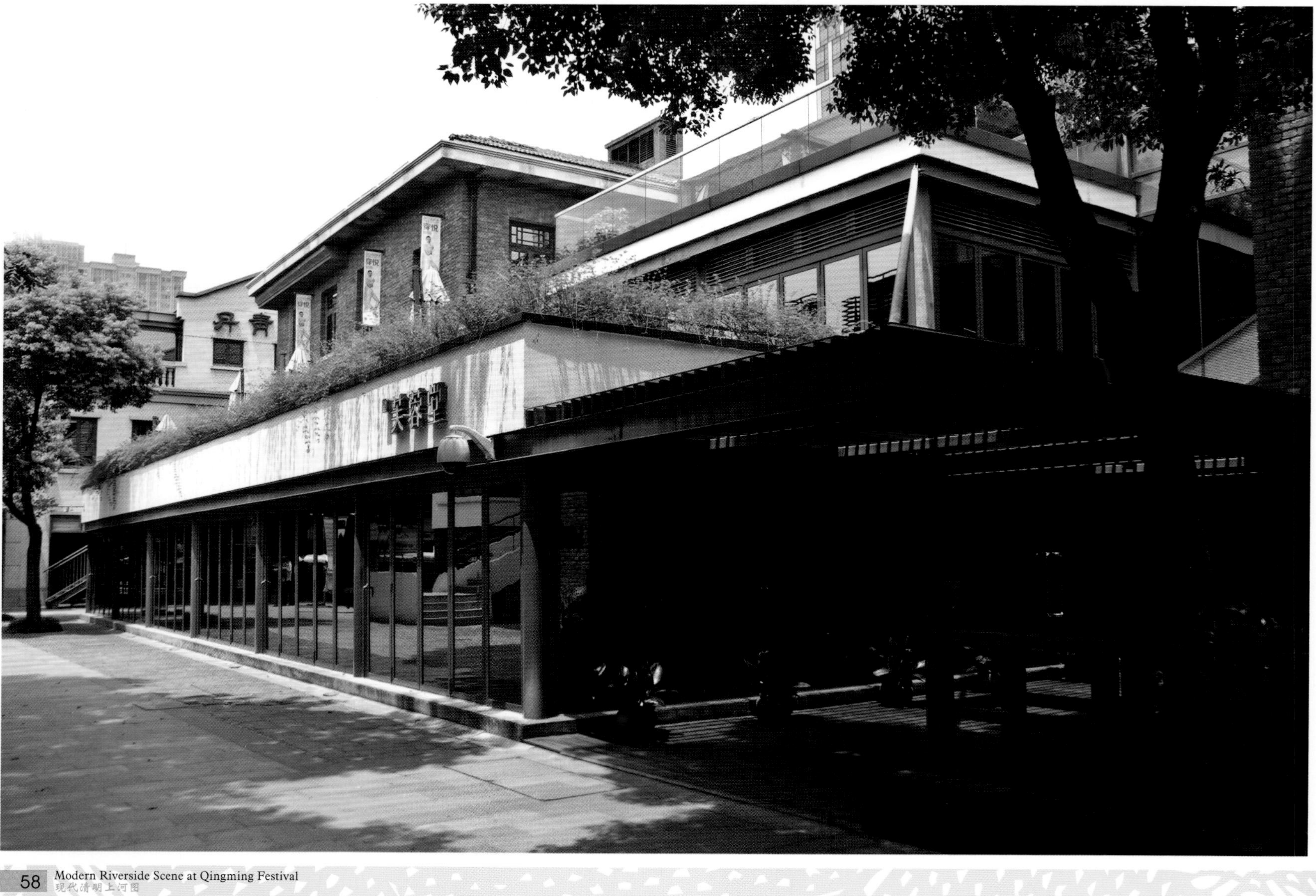

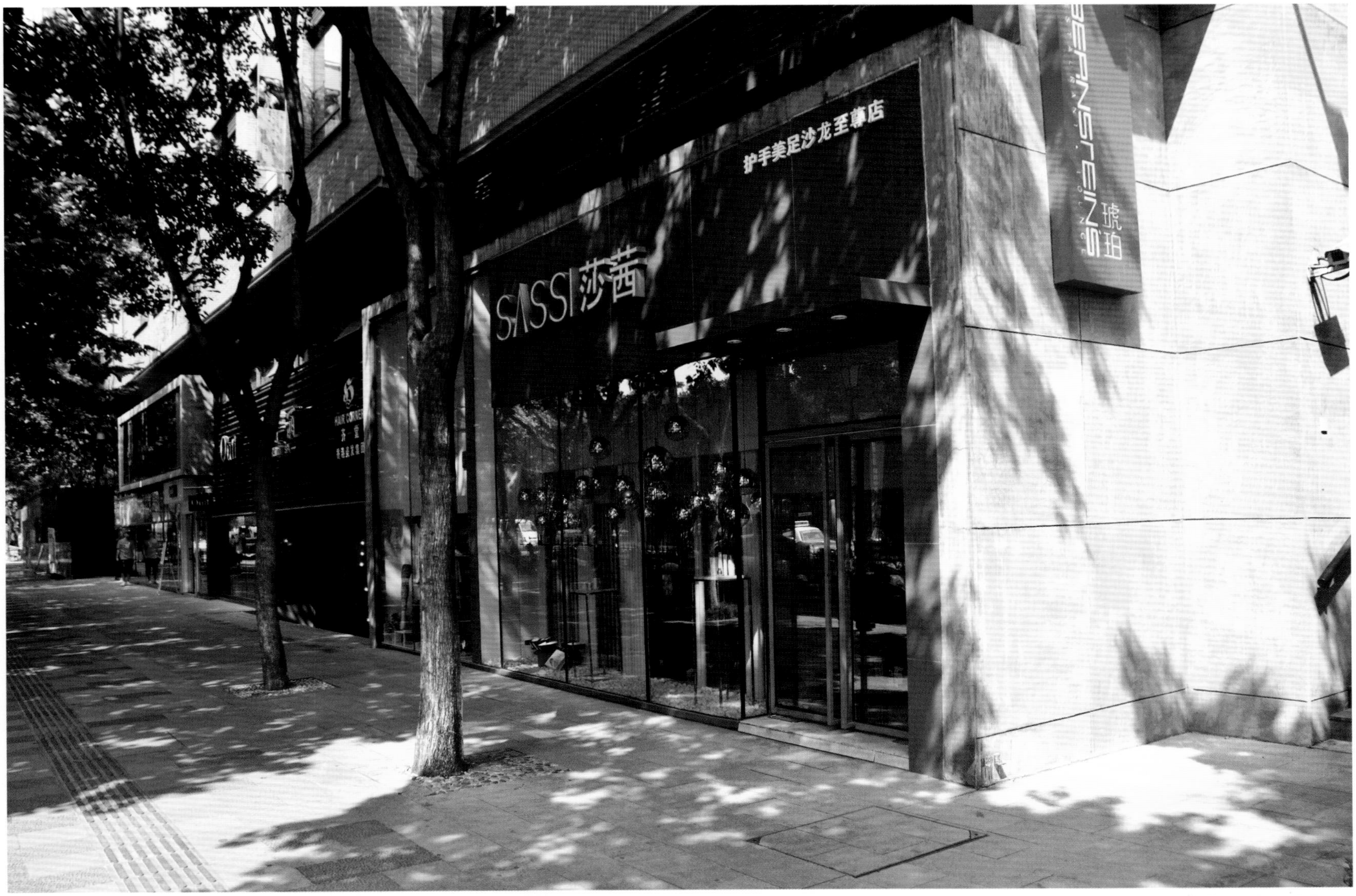
护手美足沙龙至尊店
SASSI莎茜
琥珀

V12 BAR
穿悦

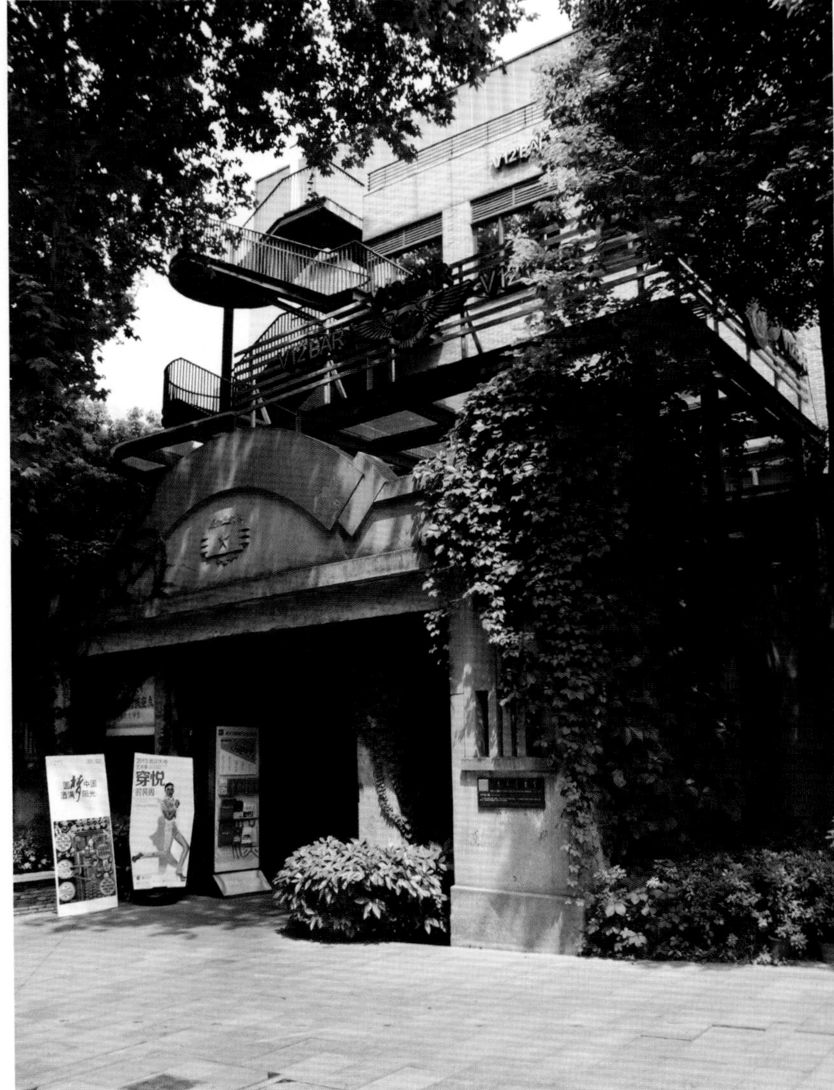

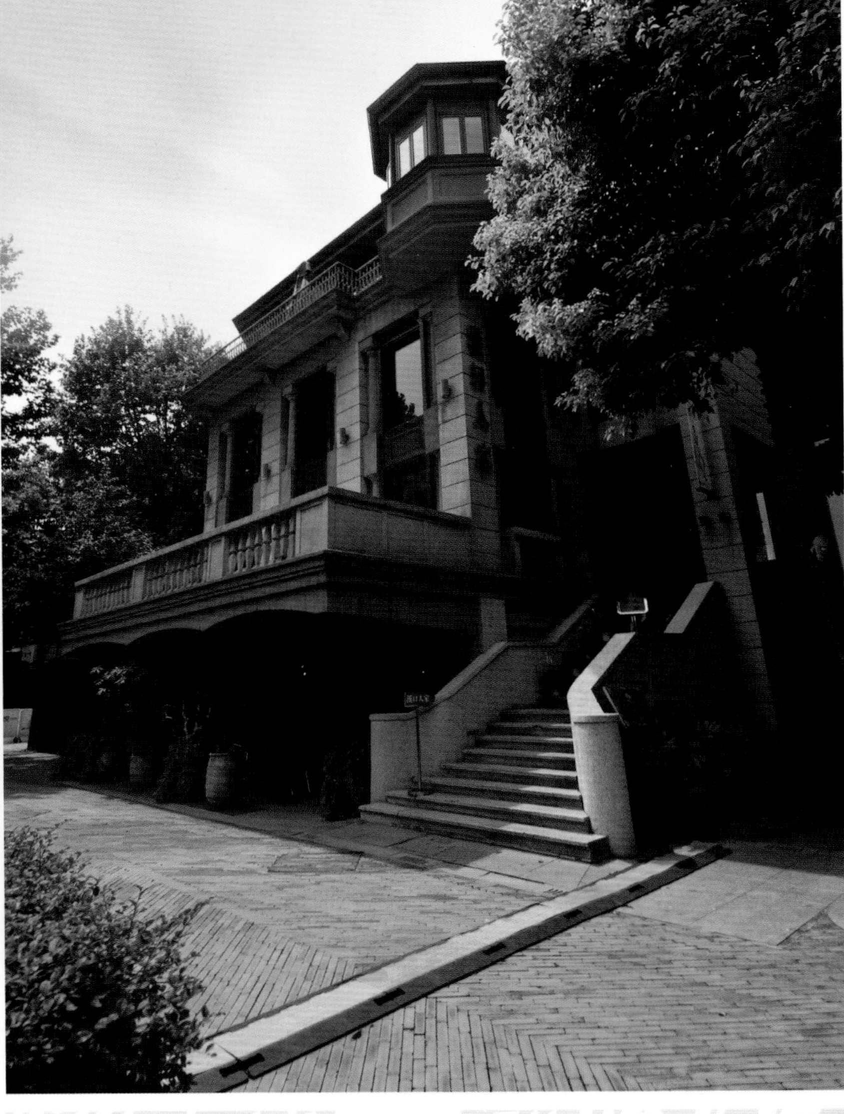

2013武汉天地
艺术季
穿悦
时装周

Majoy Commercial Activities 主要商业业态

The commercial activities in Wuhan Tiandi are mainly entertainment, restaurant and clothes among which 50% are restaurants, 20% are entertainment, 10% are clothes and 20% are others. Its merchant's occupancy rate exceeds 90%, therefore it is very prosperous.

Restaurant: Four Season Restaurant, FADO, Bernstein's Restaurant, Royal Glutton Tongue and so on.

Entertainment: Now Pub, TSRJ Hair-style Design, Beauty Farm SPA, DECLEOR Health & Beauty SPA and so on.

Clothes: Lady Embroidery, Wuhan Tiandi Gift Shop, Salut International and so on.

Others: Royal Flowers and so on.

武汉天地的主要商业业态以休闲娱乐、餐饮、服饰为主，其中餐饮占50%，休闲占20%，服饰占10%，其他商业占20%，目前商户入驻率超过九成，商业十分繁华。

餐饮业：四季恋餐厅、法朵葡园餐厅、琥珀餐厅、御甲堂养生餐厅等。

休闲业：NOW Pub、天上人间发型、美丽田园SPA、思妍丽美容护肤中心等。

服饰业：女儿绣、武汉天地专卖、日本童装、赛略特国际商店等。

其他商业：台北老爷国际花艺店等。

Food & Beverage 餐饮类

Four Season Restaurant & Jamaica Blue

Four Season Restaurant integrates natural beauty with artistic romance to create an urban space of gourmet food and music. It assembles Shanghai cuisine, Jiangsu and Zhejiang cuisine, Sichuan and Hubei cuisine, and Taiwan flavor fruit milk and ice shaving.

Jamaicablue is a coffee brand from Australia which sells authentic Jamaica Blue Mountain. Apart from coffee, it also provides hand scrolls, sandwiches, hamburgers and so on. Its design of skylight above exquisite corners makes small parties private yet not repressed.

四季恋餐厅&Jamaicablue

四季恋餐厅融自然灵秀与艺术浪漫于一体，典雅中凝透现代时尚元素，营造美食与音乐的都市独享空间，荟萃新派上海菜、苏浙菜、川鄂菜及具有台湾风味的果奶、沙冰。川鄂本色与海派技艺融汇，原味制造。

Jamaicablue是来自澳大利亚的咖啡品牌，经营正宗的牙买加蓝山咖啡。品味咖啡之余，还有手卷、三明治、汉堡等美食可供选择。透亮天窗的别致角落，使小小的聚会私密却不压抑。

Hanxiang

Hanxiang in Wuhan Tiandi mainly provides customers with middle- and high-class Korean barbecue and traditional dishes. It is operated and managed by professionals from Korea. Warm and romantic hall seats and comfortable and upscale boxes satisfy various fashionable and business consuming demands.

韩香三千里

韩香三千里武汉天地店主营中高档韩国烤肉及传统料理，餐厅由韩国专业人士经营管理。温馨浪漫的大厅卡座，舒适高档的韩式特色豪华包房，能够满足各种时尚及商务消费需求。

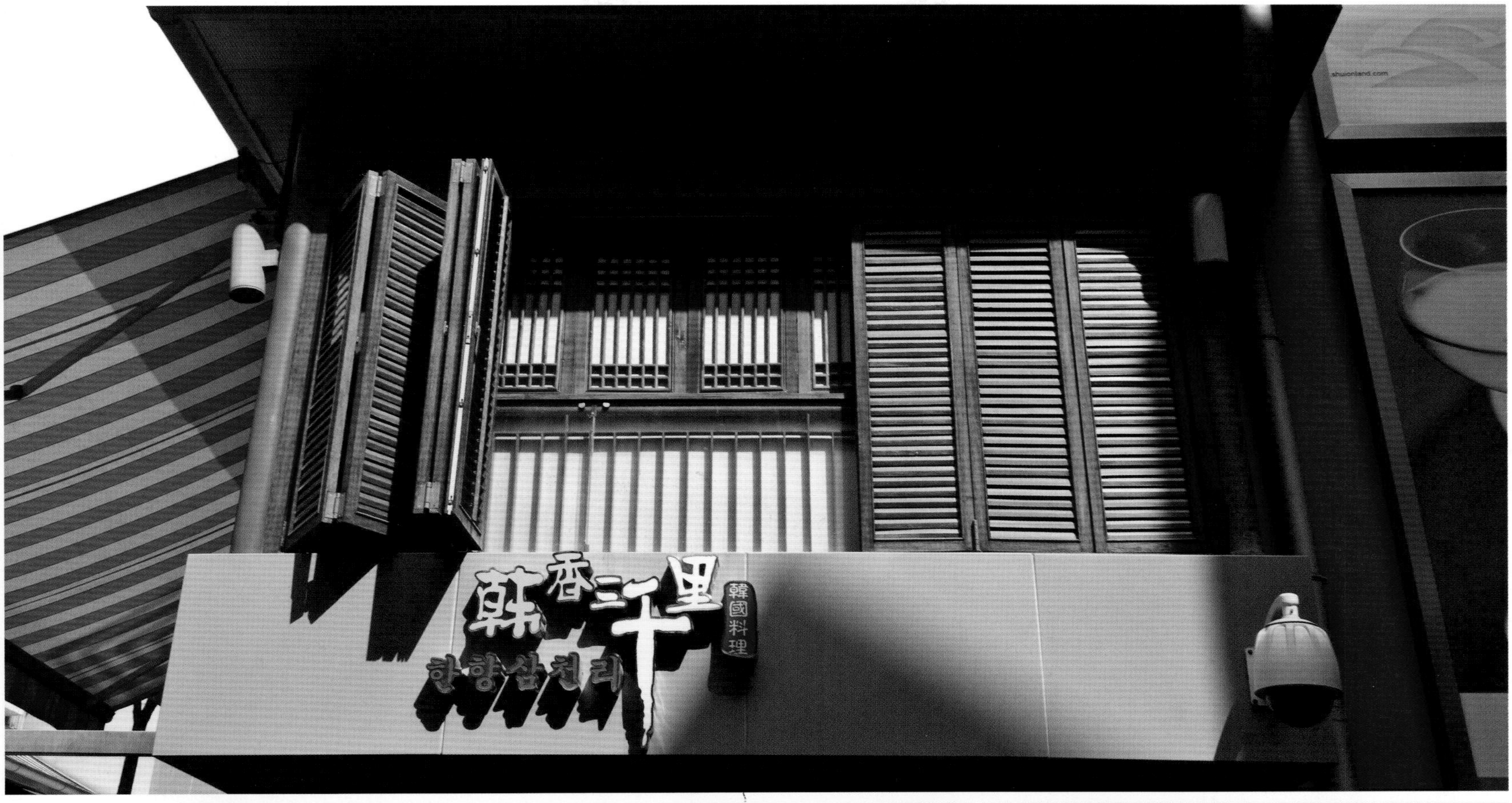

Charmant

Charmant's main product is authentic Taiwan cuisine which emphasizes on good health, never using monosodium glutamate and it is good at making sweetmeats.

小城故事

小城故事主营正宗台湾菜，饮食注重养生，做菜从不放味精，擅长做甜品。店内的菜品体现出台湾人的细致。招牌菜有三杯鸡、菠萝油条虾、台式卤肉等。

Hoi's Restaurant

Its main product is Cantonese cuisine, seafood and Hubei and Hunan cuisine. It integrates Lingnan flavor with Chu culture to present the Hoi's Restaurant's diet concept of tolerance of all styles of cuisines.

海鸣轩

海鸣轩餐厅是武汉海鸣轩酒店发展有限公司旗下的公司，主营粤菜、海鲜和湘鄂菜，把岭南风情与楚文化精髓融于一炉，菜品样式蕴含沿海风韵和湘楚特色，彰显出海鸣轩海纳百川、百家争鸣的饮食理念。

Prince Kitchen

Prince Kitchen is about 2,000 m² with totally 300 seats and 11 luxury boxes. Its lobby is over 6 meters high with a 180° circle perform stage. It is the top fashion concept restaurant in Wuhan. Its advanced and bold decoration and exquisite cuisine integrating food of Chinese, Western, Japanese, Thai style unreservedly display the charm of combination of art and cuisine.

王子厨房

王子厨房是深圳市麦广帆餐饮策划有限公司旗下一家品牌餐饮，经营面积2 000多平方米，总餐位300个，包括11个豪华包房，具有高低有致的空间和冷峻的现代设计特色。大厅有6米多高，有180度环形演艺舞台，每晚安排各种表演。它是武汉顶尖时尚的概念餐饮店，前卫大胆的装潢，融合中、西、日、泰精美的环球菜肴，尽显艺术与美食相结合的魅力。

Colorful March

The name of Colorful March originates from a ancient poetry and is a metaphor of its main product — Jiangsu and Zhejiang cuisine which are famous for their delicacy. The interior decoration of the restaurant is of simple and fashionable style which makes the restaurant elegant and warm.

品三月

品三月店名源自“烟花三月下扬州”这句古诗，暗喻主打江浙菜，其菜品以精致著称。餐厅内部是简约时尚的装修风格，淡淡的灯光、漂亮的大吊灯使整个餐厅显得典雅温馨。主要菜品有蟹粉狮子头、蒜香牛柳、扬州排骨烧肉等。

Royal Glutton Tongue & Dark Sakura

Royal Glutton Tongue features Chinese style nourishing and healthy food. Antique Chinese features, matched with modern fashionable decoration, present a post-modern classic flavor of Chinese style.

Dark Sakura's main product is Japanese cuisine. Its dinning environment is very elegant.

御甲堂&黛樱

御甲堂主要以滋补养生为卖点，推出宫廷特色小菜搭配各种养生滋补火锅，以复古的中国特色为主线加上现代时尚修饰手法，表现出中国特色的后现代古典风味。菜品主打药膳滋补甲鱼火锅，招牌菜有鲍汁甲鱼、大王蛇、八宝甲鱼等。

黛樱料理位于武汉天地3号楼一层，主营日本料理，环境优雅，推荐菜有日式软骨鸡、猪扒乌冬面、刺身、三文鱼、蟹子炒饭等。

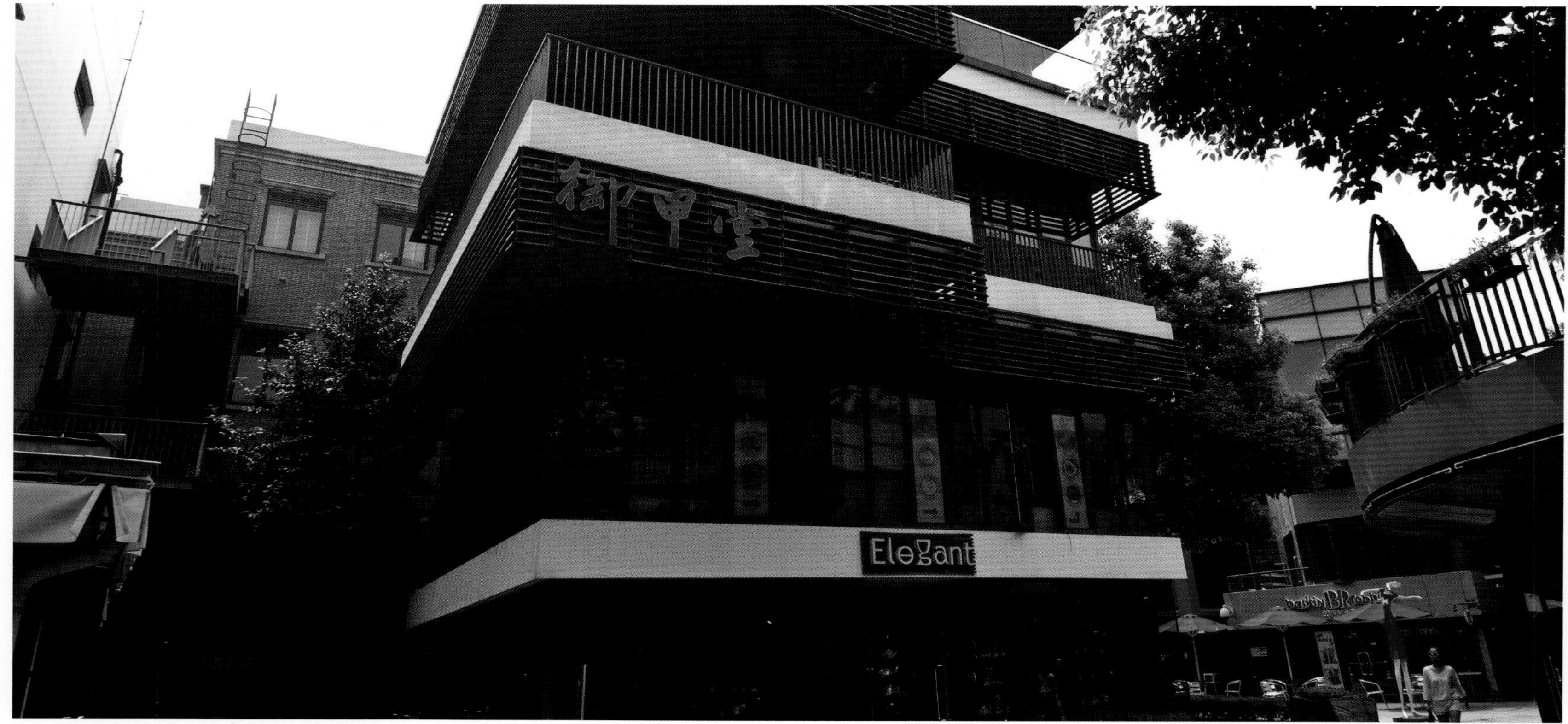

FADO & Si Chuan Cuisine

FADO inherits Portuguese pure nobility and European romanticism to create a beautiful loving environment. There are various theme parties with professional singers and colorful and shining lighting to raise customers' passion.

Si Chuan Cuisine has gathered excellent chefs and professionals and is determined to create a brand new Si Chuan cuisine. Its interior decoration is noble and elegant. It pays a great attention to material to ensure the nutritional ingredient and taste of its dishes.

法朵葡园餐厅&川源色

传承了葡萄牙纯正贵族气质和欧洲浪漫主义风情的武汉法朵葡园餐厅把含情脉脉的美丽邂逅安排在花开的时刻。专业的驻唱歌手在各类主题狂欢夜中为人们带来惊喜，魅惑的灯光透过考究的水晶吊灯流光溢彩，点燃顾客的激情。

武汉川源色管理有限公司成立于2011年，公司内部汇集了优秀厨师及餐饮界专业人士，立志开创一种全新的川菜。餐厅室内装修尊贵典雅。川源色注重选材，所用的原料都是上等鱼、虾、肉类等，保证材料原有的营养成分和口感。

Häagen-Dazs

Häagen-Dazs was born in a family of Bronx of New York in 1921. All the time, Häagen-Dazs insists on using material of top quality and pure natural ingredient to bring people an extremely tasteful feeling. For now, Häagen-Dazs has over 800 dessert houses all over the world, thus people from different places can enjoy its delicacy.

哈根达斯

哈根达斯于1921年诞生于纽约布朗克斯市的一个家庭。一直以来，哈根达斯都坚持选用最优质的配料，纯天然的成分给人们带来世上难求的味觉感受。哈根达斯发展至今，已在全球拥有800多家甜品屋，无论是奔放的美洲，还是神秘的阿拉伯世界，哈根达斯让世人都能品尝到它的美味。

Huang Ji Huang

Huang Ji Huang was founded by Huang Geng who created a unique delicious dish according to the theory of homology of medicine and food. Its operation mode has initiated several "the first" in Chinese food field and made a great contribution to realizing standardization of Chinese food, establishing healthy diet, fashion diet and so on. With its advanced operating concept and unique market expanding mode, it grew rapidly in a short time and has become a legend of chain kingdom in Chinese food field.

黄记煌

黄记煌创始人黄耕先生创造出的“黄记煌三汁焖锅”是依据“药食同源”理论创造的，源于清代御膳名肴“香辣汁鱼”，结合了传统滋补理论和现代养生学说。其操作模式开创了中餐领域多项“第一”，在实现中餐标准化，创建健康餐饮、绿色餐饮、环保餐饮、时尚餐饮等方面做出了巨大的贡献。黄记煌以超前的经营理念和独特的市场拓展方式，使品牌在短时间内迅速成长，创造了中餐领域连锁王国的一个神话。

Clothes 服饰类

Woo

Woo was founded in 2002 by Sun Qingfeng. The original intention of Woo is to present the charm and fashion of women through scarves. For over ten years, Woo has always tried to improve the inside and external beauty of urbanites who are fond of fashion, and therefore to make scarves a symbolic design treasure through their unique cultural elements.

Woo 嫵

Woo嫵创立于2002年，创始人是孙青锋先生，公司位于上海田子坊。Woo嫵创立的初衷是想通过围巾作为载体，把“静止的美变成流动的美”，让妩媚和时尚以及风情万种在女性的脖颈间流淌。10余年来Woo嫵致力于提升爱好时尚的都市男女内在和外在美的价值本身，秉承“让作品自己会说话”的设计宗旨，使其围巾和披肩通过其特有的文化元素成为具有象征意义的设计珍品。

Others 其他商铺

WUHAN TIANDI

武汉天地
WUHAN TIANDI

新四军军部旧址
Memorial Site of New Fourth Army Headquarters
八路军办事处旧址
Memorial Site of Eighth Route Army in Wuhan
200m
武汉天地
"Wuhan Tiandi" Commercial Area

敬请取阅
穿悦

Xintiandi, Shanghai
上海新天地

街区背景与定位
Street Background & Market Positioning

History 历史承袭

Xintiandi, Shanghai, was rebuilt on the base of the old residential area of Shanghai Gate whose history dated back to 1860s . It was Taiping Rebellion when a large number of refugees came to Shanghai Concession for refuge. Merchants were mobilized to invest to build residences for refugees. The style of these residences was similar to the style of Jiangnan residences.

Buildings in Shanghai Gate have a long history. They are the combination of Chinese and Western elements, but also the witnesses of Shanghai-style culture. With the change of family structure and people's residential concept, the Shanghai Gate, though used to be bustling, couldn't satisfy the living demands of residents any more. In 1997, Shui On Group changed the original residential function of Shanghai Gate, and gave it a function of commercial operation; therefore, the Xintiandi was created.

上海新天地是在石库门建筑住宅旧区的基础上改造而成的，住宅区历史可追溯至19世纪60年代。时值太平天国运动，大量难民为躲避战乱涌入上海租界避难，为接纳难民，租界动员商人投资新建了联排式的石库门里弄住宅。这些住宅风格与江南普通民居相似，大部分分布在河南中路东侧、兴仁里及中山南路新码头街的敦仁里、棉阳里、吉祥里等地。

石库门建筑有着深深的历史烙印，它既是中西合璧的产物，又是海派文化的见证。随着家庭结构的变化和人们居住理念的改变，昔日热闹非凡的石库门早已不能满足人们的居住需求，黯然地淡出历史舞台，甚至一度有专家忧郁地哀叹："上海将见不到原汁原味的石库门了。"1997年，瑞安集团在石库门建筑住宅区的基础上，改变石库门原有的居住功能，赋予它商业经营的功能，石库门获得新生，上海新天地由此而来。

Location 区位特征

Xintiandi is located at No.181, Taicang Road, Huangpu District, on the south side of Huaihai Zhong Road where is between South Huangpi Road and Madang Road. Situated at the city center, the most prosperous area of Shanghai, Xintiandi is occupying abundant tourism resources.

新天地位于上海市中心，在淮海中路的南侧，处于黄陂南路和马当路之间，毗邻黄陂南路地铁站和南北、东西高架路的交会点，落户于上海市黄浦区太仓路181弄。新天地处于上海最繁华的地段，其周边有众多的旅游景区，旅游资源丰富。

Market Positioning 市场定位

Xintiandi is a tourist attraction of Shanghai historic culture, being both traditional and modern. After transforming Shanghai Gate, it has become a fashionable entertaining culture center which assembles international food, shopping and entertainment.

The market positioning has been deepened for three times according to different development phases of Xintiandi since 1997. The first phase put emphasis on comprehensiveness which means to integrate food, entertainment, shopping, tourism and culture. And its focal area is Hengshan Road. The second phase aimed to make Xintiandi be a urban tourism attraction with historic culture feature. The third phase dedicated to building an international spot for communication and gathering.

上海新天地是一个具有上海历史文化风貌、中西融合的、充满传统和现代气息的旅游景点。在改造石库门老建筑之后，它成为了集国际水平的餐饮、购物、演艺等功能于一身的时尚休闲文化娱乐中心。

自1997年石库门老建筑改造开始，上海新天地依据其开发程度的不同，定位也先后进行了三次深化：第一阶段定位强调综合性，即集餐饮、娱乐、购物、旅游、文化于一身，此阶段开发的重点区域为衡山路；第二阶段定位是将上海新天地打造成为具历史文化特色的都市旅游景点；第三阶段定位是将上海新天地打造成一个进行国际交流和聚会的地点。

Street Planning 街区规划

1. Planning Functional Area

Xintiandi is divided into south part and north part, which represent modern fashion and traditional charm. Combining the boundary of north part and south part, sideward greenbelt and lakes, the functional area of Xintiandi is planned with four parts.

1) Nan Li, (the south part). Centralizing shopping, entertainment and leisure, there is a building with glass curtain wall of modern breath. It contains restaurants from all over the world, and fashion exclusive shops, accessory shops and a cinema that favored by youths.

2) Bei Li, (the north part). Several old buildings of Shanghai Gate are integrated with newly-built modern buildings and modern interior decoration and equipment to create many upscale shops and restaurants.

3) Xingye Road. This is the boundary of north part and south part, located at the site of the First Congress of the CPC. The site together with other buildings along the street composes a scenery line with rich historical culture.

4) Park greenbelt and artificial lake. Park greenbelt, which has a large area, is located at the center area of Taiping Bridge Project. There are verdant trees and resting space in the park. The biggest artificial lake in Shanghai is situated at the middle of the park. The lake is decorated with fountain, Yulan Island at the east and Hehuan Island at the west, and trees and shrubs around the lake. A lakeshore path is added to the north along the lake to make the scene of Xintiandi more wonderful.

2. Specific Street Planning

According to the situation of Shanghai Gate, Shui On Group decided to implant the old buildings with fresh energy from the perspectives of protecting historical buildings, city development and building functions. Specific planning is as follow.

The total planning reserves most old buildings in the north part and alternates them with some modern buildings. The south part is mainly newly-built modern buildings with seldom old buildings. These two parts are linked by a pedestrian street.

In the premise of protecting the feature of plain brick wall and historic sense of Shanghai Gate buildings, designers kept the original bricks and tiles as materials. On the base of reserving the old buildings' appearance, designers planned the interior structures and functions and added modern equipment to the buildings to ensure the functions of these buildings.

1.规划功能区

上海新天地分南里和北里两部分，分别代表了现代的时尚和传统的韵味，结合新天地南、北里的分界和旁边的绿地、湖泊等场所，新天地功能区规划有以下四部分。

1） 南里。以购物、娱乐、休闲为中心，这里建成了一座充满现代气息的玻璃幕墙建筑，面积达25 000平方米，引入各具特色的商户，设有来自世界各地的餐饮店，同时还囊括了年轻人喜爱的时装专卖店、时尚饰品店、电影院等。

2） 北里。由多栋石库门老建筑组成，结合新建的现代建筑、室内现代的装潢和设备，成为多家高级消费场所和餐厅。

3） 兴业路。这是南里和北里的分水岭，中共“一大”会址坐落其间，会址与沿街石库门建筑共同组成一条历史文化浓厚的风景线。

4） 公园绿地与人工湖。公园绿地占地面积极广，处于整个太平桥项目的中央地带，园内树木葱郁，提供休憩空间，公园中心修建了上海最大的人工湖，湖中修建喷泉，东西以“玉兰岛”和“合欢岛”点缀，湖四周饰以乔木和灌木。绿地北侧沿人工湖修建一条湖滨路，流畅的曲线与湖西的石库门海派建筑融为一体，为新天地更添一抹风景。

2.街区的具体规划

针对石库门老建筑的现状，瑞安集团从保护历史建筑、城市发展以及建筑功能等多角度考虑，决定要为这些旧建筑注入新生命以满足消费者需求。具体规划如下。

在整体规划上保留了北部地块大部分的石库门建筑，穿插部分现代建筑；石库门南部地块则以新建现代建筑为主，配合少量石库门建筑；南、北两个地块以一条步行街串连。

在保护石库门建筑清水砖墙的特色及其历史感的前提下，设计师保留原有的砖、瓦。在保留建筑外衣的基础上，对建筑内部结构和功能进行规划，加装现代设施，诸如地底光纤、空调系统等，确保房屋功能完善、可靠。

Street Design Features 街区设计特色

Xintiandi keeps the original brick walls, roofing tiles and building group of Shanghai Gate to provide visitors with a sense of placing themselves in the living environment of old Shanghai in 1920s. However, the interior of every building is decorated according to the modern urban life to form international galleries, fashion shops, theme restaurants, cafes and bars. Xintiandi successfully integrates traditional charm with modern fashion and becomes a best place for tourists to appreciate historic culture and modern life of Shanghai, as well as a gathering place for local citizens and foreigners.

新天地保留了当年的砖墙、屋瓦、石库门建筑群，让游客穿越时空，置身于20世纪20年代老上海的生活环境之中。但新天地每座建筑内部，又按照21世纪现代都市人的生活方式、生活节奏、情感世界量身定做，成为国际画廊、时装店、主题餐馆、咖啡馆、酒吧等。新天地成功地将传统的风韵与现代时尚的生活融为一体，成为中外游客领略上海历史文化和现代生活的最佳去处之一，也成为本地市民和外籍人士的聚会场所。

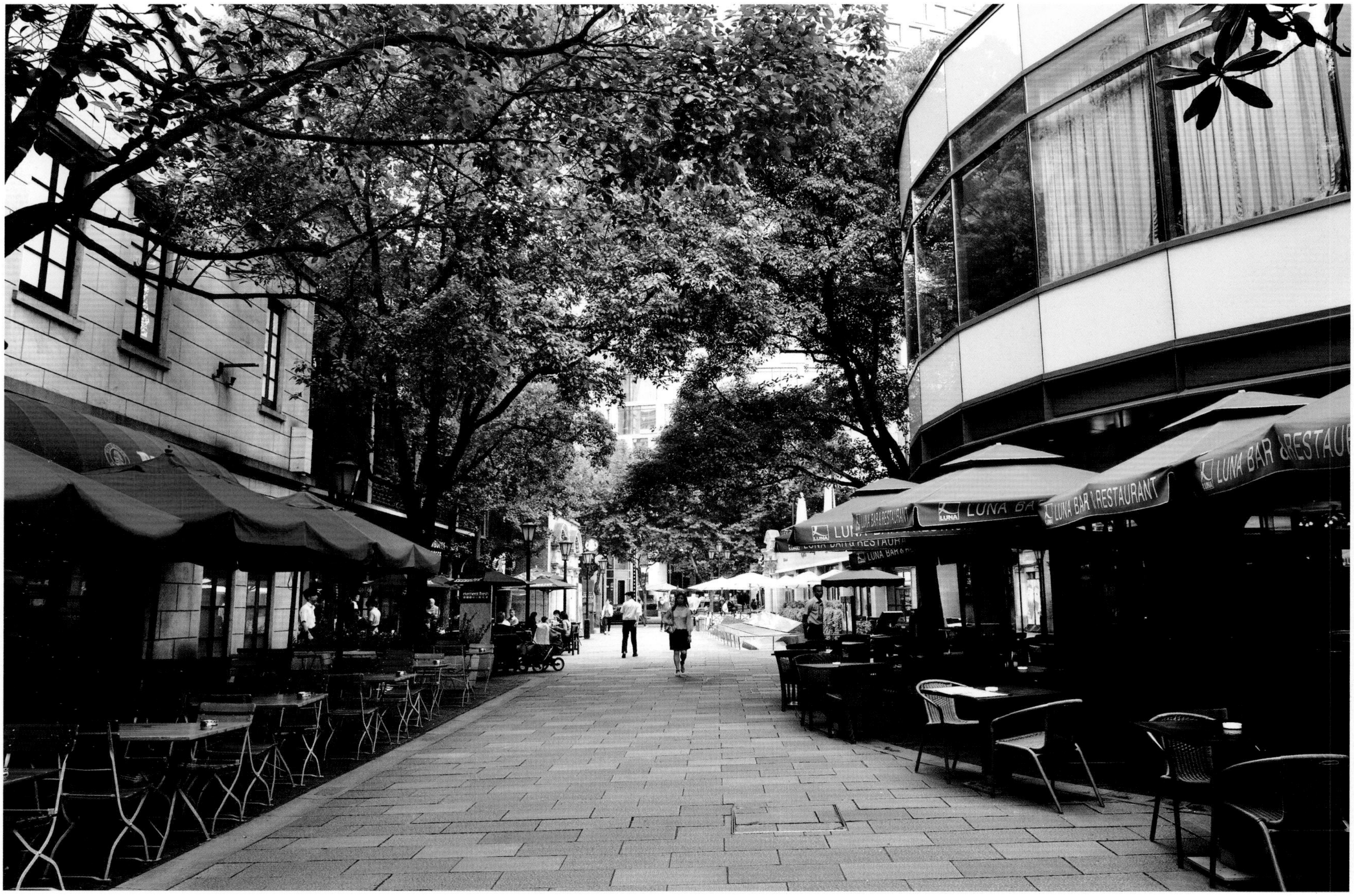
LUNA BAR RESTAURANT

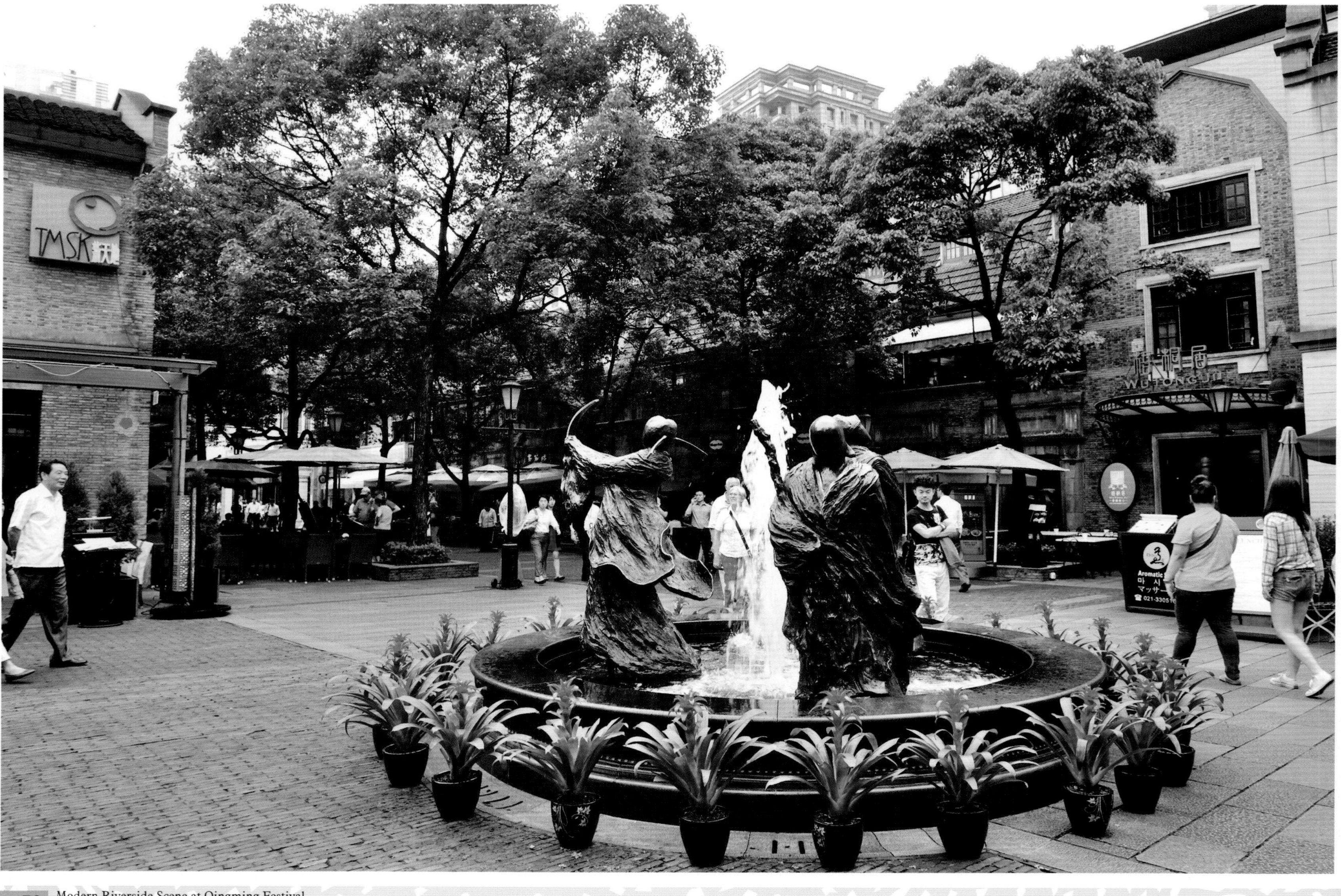
TMSK

石库门屋
SHIKUMEN OPEN

SHANGHAI TANG

鈺
ZEN
lifestore
鈺
ZEN
lifestore

屋里厢
茶馆

天泰
simply thai

T8

KABB
ye shanghai

商业业态

Commercial Activities

After transforming old buildings in Shanghai Gate, Xintiandi has become a fashionable leisure & culture center which assembles international food, shopping and entertainment. Its shop renters are all worldwide well-known brands with 85% of them from countries and regions other than Chinese mainland. Its distribution of commercial activities is mainly as follows.

Nan Li: it gathers gourmet restaurants of different countries and regions, fashion exclusive shops, fashionable accessory shops, cinemas and fitness centers to provide consumers with multiple and tasteful leisure entertainment.

Bei Li: leading by the concept of "integrating the new and the old, combining Chinese and Western elements", there are many upscale restaurants offering different dishes around the world, elegant luxury shops and bars.

上海新天地在改造石库门老建筑之后，成为了集国际水平的餐饮、购物、娱乐等功能于一身的时尚休闲文化中心。其招租的对象均是来自世界各地的知名品牌，85%左右的租户来自中国内地以外的国家和地区。其商业业态分布也与南里、北里地区分布有关，主要形成了如下分布。

南里：云集各国美食餐厅、时装专卖店、时尚饰品店、电影院、健身中心等，为消费者及游人提供多元且极具品位的休闲娱乐。

北里：在“新旧交融，中西合璧”理念的指引下，北里建有多家高级消费场所、来自世界各地的餐厅、精致典雅的名品店及酒吧。

Food & Beverage 餐饮类

Brown Sugar

Brown Sugar is a well-known Jazz bar in Taipei. Brown Sugar of Shanghai is located at the Bei Li Square of Xintiandi with over 200 seats to give consumers a new experience of appreciating Jazz music. It is common to see stars in Brown Sugar. Foods, which are mostly European style in Brown Sugar, are also very delicious.

Brown Sugar

Brown Sugar是台北知名的爵士酒吧。上海的Brown Sugar位于新天地北里广场，有两百多个座位，最多可容纳五百人，给人一种全新的爵士乐观赏体验。在Brown Sugar里面可以经常见到明星，奠定了新天地新音乐地标的地位。Brown Sugar内的美食也是一绝，菜系以欧陆菜色为主。

TMSK

TMSK situates at Bei Li Square. It has won lots of honors. *The Times* compared TMSK to a treasure in April 2009, and listed it as one of recommended restaurants of Shanghai EXPO. The interior of TMSK is extremely luxurious. Consumers can appreciate glass made of colored glaze freely and even taste the wine or beverage in them.

TMSK

TMSK（透明思考）位于新天地广场北里，这家餐厅有着诸多荣誉。2009年4月，英国《泰晤士报》将TMSK餐厅比作一块珍宝，2010年，其被列为Shanghai EXPO推荐餐厅之一。餐厅内部异常的奢华，琉璃制成的杯子任人把玩，人们还能品尝到里面盛放的酒与饮料。

Starbucks

Starbucks founded in 1971 is the largest coffee chain shop in the world. Apart from coffee, there is tea, pies and cakes in Starbucks. The Starbucks in Xintiandi has a unique space design together with high quality coffee and service, providing consumers with a good place for tasting coffee.

星巴克

星巴克是美国一家连锁咖啡公司，1971年成立，是全球最大的咖啡连锁店。除了咖啡外，星巴克还有茶、馅饼及蛋糕等。上海新天地中的星巴克咖啡是上海统一星巴克有限公司的分店，店内独特的空间设计理念，配以高品质的咖啡与服务，为消费者提供一个品尝咖啡的好去处。

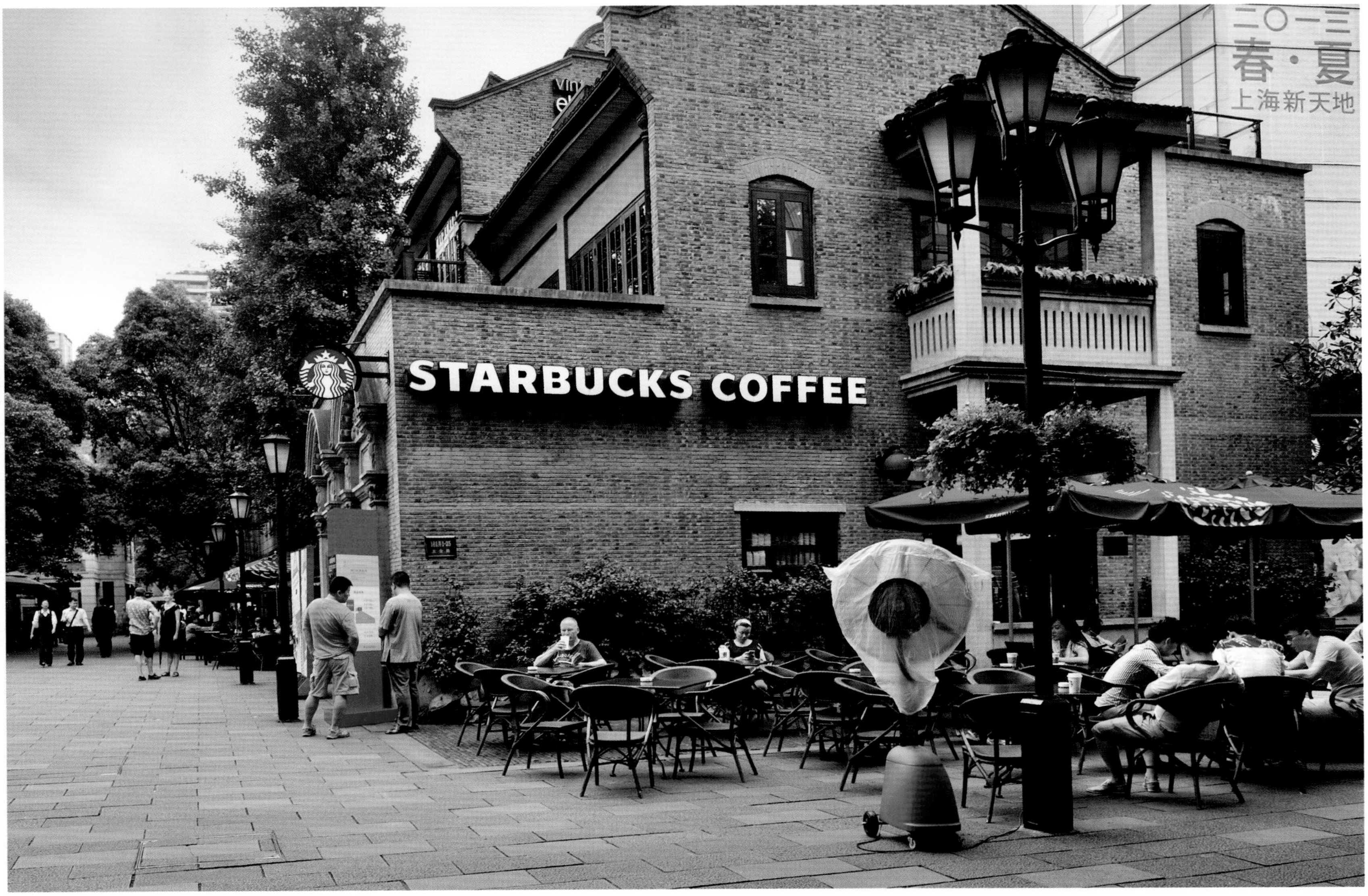

The Coffee Bean & Tea Leaf

The Coffee Bean & Tea Leaf in Xintiandi though is not big, its decoration is exquisite and comfortable. The light music in the shop adds quietness and romance to it. It is a good place for relaxation.

香啡缤

香啡缤（新天地店）位于北里10号楼，地方不大，布置十分精致、舒适。店内播放着轻柔的音乐，显得幽静而浪漫。店内咖啡不错，是休闲的好去处。

ZEN Restaurant

ZEN Restaurant in Nan Li is decorated with design elements of the Buddhism such as smooth arcs, continuous white walls and lotus lamps to provide consumers with a quiet and peaceful feeling. Its dishes stress reasonable and healthy collocation of nutritional ingredient and are changed with the seasons. Meanwhile, its art of serving the dishes, including the dishing up, the expression of appearance, aroma and taste, and especially the serving of dishes at the right time while at the most proper temperature, is world-famous.

采蝶轩

采蝶轩位于新天地南里。步入采蝶轩，入眼的是光滑的弧线、绵延的白墙、莲花灯等带有佛学意境的设计元素，使人心灵深处不自觉地升起一种静谧祥和的感觉。在一个神秘、肃穆，同时又不失灵动感的空间里品尝传统精致的中国菜，真的有世外桃源的感觉。另外，它的菜肴讲究合理健康的营养成分搭配，并随着季节的变化而有应时的菜式变化，同时还以上菜的艺术（包括每道菜的装盘、色香味的传达、尤其是选择菜的热度恰到好处的那一刻上菜）而闻名中外饮食界。

Simply Thai

Simply Thai which belongs to Simply The Group is a Thai restaurant. It's a perfect combination of Asian tradition and modern concept. The branch of Simply Thai in Xintiandi is not big, with two floors of nostalgic Shanghai Gate house and open-air seat outside. Its featured dishes include green curry, coconut sago cake, Thai swamp cabbage,etc.

Simply Thai

Simply Thai属于Simply The Group公司旗下成功的餐饮品牌，是一家泰国餐厅。这一品牌餐厅正在定义“新中国”生活方式，是亚洲传统和现代理念的完美结合。Simply Thai在新天地的分店不大，上下两层，石库门的房子有点怀旧，门口设有露天座位。它的特色菜品有绿咖喱、青咖喱、椰子西米糕、泰式空心菜等。

Fountain

Fountain is a comparably distinctive Western restaurant in Xintiandi. The chef of the restaurant, who is good at various Western foods including British style, French style and Italian style, used to be a cook employed by Lee Kuan Yew, the former prime minister of Singapore.

丰泉

丰泉是上海新天地里比较有特色的西餐厅，餐厅原是以前的“逸飞之家”。餐厅的主厨曾经是新加坡前总理李光耀的厨师，擅长英式、法式、意式等多种西餐。另外，丰泉的早餐和下午茶也有多种口味，招牌甜品“百宝箱”更是一绝。

VABENE

VABENE is an Italian restaurant from Hong Kong. It is two-floor with decorations of wooden flooring, slate, wooden window frames, glass wine cabinet and romantic Tuscany color. The first floor is a glasshouse in which customers can enjoy natural light when tasting delicious food. The second floor is a bar decorated with fireplace and wooden column, which is totally different from the first floor.It provides mainly Italia-style food.

VABENE

VABENE是一家来自香港的意大利餐厅，餐厅有两层，原木地板、铺石板、木窗框、玻璃酒柜以及浪漫温和的托斯卡尼色调，很有昔日上海颓废的美感。餐厅一层是装有玻璃的温室型用膳环境，可供客人一边享受天然日光，一边品尝美食；二层是以原石壁炉和木柱装饰的酒吧，呈现出与一层截然不同的感觉。餐厅主营意大利美食。

Lind Bakery

Lind Bakery is a German family-run bake enterprise of over one hundred years' history. Its main products are traditional German-style bread including soda bread, grain bread and so on.

琳德烘焙

琳德是一个有百余年历史的德国家族烘焙企业，由老罗伯特·琳德于1905年在一个德国小镇建立。在一个多世纪里，生意一代一代传承并发展壮大。企业主营传统正宗的德式面包，有碱水面包、粗粮面包等多个品种。

South Beauty

After dozens years of development, South Beauty has become one of the most promising international food & beverage service management companies and is leading Chinese food culture towards international market.

俏江南

俏江南是俏江南股份有限公司旗下的餐饮品牌之一，于2000年由张兰女士创立。历经十余年的发展，俏江南已经成为中国最具发展潜力的国际餐饮服务管理公司之一，并引领着中华美食文化走向国际市场。

Xin Ji Shi Restaurant

It is the first client that enters Xintiandi and also the only one restaurant operated by Shanghailander. It locates at a residence of Shanghai Gate built in 1925 and shows a strong Shanghai flavor. Its dishes are mainly local cuisine and its customers are mainly foreigners.

新吉士酒楼

新吉士酒楼是第一家入驻新天地的商户，也是唯一一家由上海人经营的餐厅。新吉士位于一栋1925年的石库门里弄住宅，门口的小天井、乌漆厚木大门、门口的大红灯笼，均展露出浓浓的上海风情。酒楼的菜式主要是本帮菜，酒楼的顾客大多是外国人。

Really Good Sea Food

Really Good Sea Food is from Taiwan. It has a good environment with its exquisite interior decoration which provides a Chinese idyllic feeling.

真的好

真的好海鲜餐厅来自台湾，餐厅环境非常好，青红相间的砖墙和雕着巴洛克风格的门楣、墙上挂着的中国清末的仕女图为餐厅平添几分文人墨客的诗情画意。推荐菜有法式焗明虾、烧汁杏鲍菇、黑椒嫩菲力、香酥鲜虾卷等。

TWININGS

Twinings is founded by a British Thomas Twining in 1706 and it was firstly named Thomas. In 1837, it was appointed as special tea for royal household by Queen Victoria. Now it is popular in over 100 countries. Apart from traditional black tea and green tea, it also provides fruit tea, flower tea and ice tea, etc.

TWININGS

TWININGS是由英国人托马斯•川宁创立的品牌，1706年川宁以Thomas之名在英国开设的咖啡馆是TWININGS的前身。1837年，英国维多利亚女王颁发第一张“皇室委任状”，川宁茶被指定为皇室御用茶。现在川宁已风靡全球100多个国家，川宁茶的品种除了传统的红茶、绿茶系列外，还有果茶、花茶、冰茶等系列。

Clothes 服饰类

Shanghai Tang

Shanghai Tang is a luxury fashionable brand. The clothes sold in this shop are low-key yet decorous, simple yet fashionable. Due to its good service manner, it gains lots of praises from customers.

上海滩

上海滩是中国奢侈时尚的先锋品牌，其新天地店位于新天地北里15号。店内服饰低调不失典雅，古朴不乏时尚，店内服务员态度很好，顾客赞誉颇多。

Club 会所类

One Xintiandi

One Xintiandi, situated at No.1 building of Bei Li, is reconstructed from a four-floor composite structure building. Its appearance is of Shanghai style folk residence, which is rare in old residences of Shanghai nowadays.

The first floor of the club is often used for painting and calligraphy display and the second and third floor are restaurants which are suitable for business dinners and private parties, for their quiet and comfortable environment. The dishes of the restaurants are very unique, combining Chinese and Western elements and are very delicious.

壹号会所

壹号会所（新天地店）位于北里1号楼，由原有的一栋四层混合结构建筑改造而成。但此建筑原先不属于石库门的房子，它的外形为“弄堂公馆”式建筑风格，目前在上海的老式住宅中已经相当罕见了。

会所一楼常年举办书画展，二、三楼是餐厅，比较适合商务宴请和私人宴会，环境幽雅，墙壁上挂着装饰画，古老的荷花灯和钟摆、时代久远的茶几和梳妆台表现出主人的品位。菜肴也很有特色，中西合璧，非常美味。

Craft 工艺类

Blancpain

Blancpain, founded in 1735, is the oldest existing wristwatch brand. Blancpain produces the most complex manual mechanical watches with fully functions in the world. Without flow process, its manufacturing process is elaborately conducted in old farmhouses by watchmakers. It never produces quartz watches, and its all products are masterworks.

宝珀

宝珀创于1735年，是现存历史最悠久、最古老的腕表品牌。宝珀手表是目前世界上最复杂、功能最多的全手工机械表。它没有流水作业式的工厂，制造过程全部在古旧的农舍内进行，由个别制表师精工镶嵌。宝珀从不生产石英表，其腕表系列中所蕴含的机动系统繁复精巧，每一款都是制表艺术中的极品，主要款式有超薄表、月相盈亏表、万年历表、双码表及三问表等。

Zen Life Store

Zen Life Store was founded in 2006 by Wang Zhengyun. It starts with selling rich and delicate museum art gifts, and devotes to spreading world museum culture, thus to enable more people to feel the human's civilization more visually.

钲艺廊

钲艺廊创建于2006年，创始人是王钲云。2012年，钲艺廊旗下新店“GIFT ZEN•钲礼”在上海新天地开业，获得了多家世界著名博物馆的授权，从丰富而精美的博物馆艺术礼品入手，致力于传播世界博物馆文化，让更多人可以更直观地感受人类文明的灿烂精华和辉煌记忆。

Jewelry 珠宝类

CARAT

CARAT was founded in 2003 by a British, Scott Thompson. The prices of its commodities vary from several thousands to tens of thousands RMB, and are affordable to general people. Now it has dozens of exclusive shops all over the world including London, Hong Kong, Dubai and so on.

CARAT

CARAT于2003年由英国人Scott Thompson创办，其商品价格从几千元到上万元不等，是一般民众都能消费得起的价位，现在CARAT在全球已经拥有了几十家专卖店，遍布伦敦、香港、上海、迪拜等城市。

Culture Facilities 文化设施

The Site of the First Congress of the CPC 中共一大会址

The site of the First Congress of the CPC is located at No.76, No.78, Xingye Road, which is the boundary of Nan Li and Bei Li. It has two two-floor brick-wood structure buildings side by side along the road and facing south. These two buildings built in 1920 are Shanghai style old buildings of Shanghai Gate. The interior arrangements are kept as original and the museum is equipped with three display rooms to show cultural relics of founding time of CPC. After being expanded in 1996, the museum's site area is 715 m^2, with tourist service facilities on the first floor and exhibition hall of cultural relics on the second floor. There are almost forty thousand cultural relics collected in the museum.

中共一大会址位于上海兴业路76、78号，处在上海新天地南里和北里的分界线上，是沿街并排的两幢两层砖木结构建筑，坐北朝南。会址建于民国9年（1920年）夏秋之际，是具有上海地方风格的石库门楼房，外墙青红砖交错，镶嵌白色粉线，门楣有矾红色雕花，黑漆大门上配铜环，门框围以米黄色石条。会址的室内布置维持当年的原样，纪念馆内设有三个陈列室展示中国共产党创立时期的文物。1996年会址纪念馆扩建，新建筑的外貌与原会址建筑相仿，占地面积715平方米。一楼为观众服务设施，设有门厅、多功能学术报告厅和贵宾厅。二楼为历史文物展览厅，馆内共收藏有文物近4万件。

Qianmen Street, Beijing
北京前门大街

街区背景与定位
Street Background & Market Positioning

History 历史承袭

In the middle of Ming Dynasty, due to development of business, street markets like Xianyukou, Zhushikou, Meishikou appeared at Qianmen Street, thus Qianmen Street became a commercial street. After Jiajing Period of Ming Dynasty, officials set up guilds to solve the accommodation problem of candidates for the imperial examinations from around the country. Candidates often came to purchase articles for daily use or to relax themselves, therefore, Qianmen Street was prompted to be a prosperous commercial street.

In the early Qing Dynasty, in order to maintain the dignity of imperial power, businesses such as theatres and teahouses were allowed to be opened only outside the city, thus Qianmen Street was further developed. Many specific street markets were opened at both sides of the street successively in Qing Dynasty. Sheds were gradually reconstructed into bricks and wooden houses. Until the late Qing Dynasty, there was night market on Qianmen Street.

前门大街在明、清至民国时期皆称正阳门大街，民众俗称前门大街，1965年正式定名为前门大街。

明朝中叶，由于商业发达，前门大街两侧出现了鲜鱼口、猪（珠）市口、煤市口等集市和街道，成为一条商业街。明嘉靖以后，各省市在京做官的人为了解决进京应试举子的住宿问题，在前门大街两厢建立了各地会馆。举人们常到前门大街购买生活用品或饮酒作乐，促使前门大街成为一条繁华的商业街道。

清初，东城的灯市挪到前门一带，为了维护皇权的尊严，戏院、茶园等只准许开设在城外，于是前门大街较前又有了进一步的繁荣。大街两侧陆续形成了许多专业集市，如鲜鱼市、肉市、果子市、布市、草市、猪市、粮食市、珠宝市、瓜子市等。附近胡同内随之出现了许多工匠作坊、货栈、车马店、旅店、会馆以及庆乐、三庆、华乐等戏园。大街的席棚之房逐渐改建成砖木结构的正式房，形成了东、西侧房后有里街的三条街。至清末，前门大街已有夜市。

Location 区位特征

Qianmen Street, as a well-known commercial street in Beijing, is located at the axis of Beijing extending from Moon Bay at the north to Tianqiao Road entrance at the south. It used to be a appointed road for emperors when they went to Temple of Heaven.

The street almost completely preserves the traditional lanes, city texture, space form and historical buildings, therefore it has a high historic culture value.

前门大街是北京著名的商业街，位于京城中轴线，北起前门月亮湾，南至天桥路口，与天桥南大街相连。明嘉靖二十九年（1550年）建外城前，其是皇帝出城赴天坛、山川坛的御路，建外城后为外城主要的南北街道。

区域内较为完整地保存了传统街巷、城市肌理、空间形态和众多历史建筑，具有较高的历史文化价值。建筑以一层为主，局部为二层。

Market Positioning 市场定位

Themed by historic culture of ancient capital, Qianmen Street integrates functions of culture, food, entertainment, leisure, fitness, shopping, business and tourism. It is positioned as a historic cultural tourist commercial street targeting at the nation and the world.

以古都历史文化为主题，融文化、美食、娱乐、休闲、健身、购物、商务、旅游多功能为一体，面向全国、全世界的历史文化旅游商业区，打造古老与现代、传统与时尚、民族品牌与国际品牌多元并生的现代街区形象。

规划设计特色
Planning & Design Features

Street Planning 街区规划

The planning principle is "taking protection as Key point, highlighting focal point, organic integration, emphasizing innovation, elaborate construction and sustainable development".

Buildings at both sides of the street are respectively managed in four ways: keeping, repairing, reconstructing and renewing. And according to four streets and 22 hutongs adjacent to it, Qianmen Street is divided into five function areas.

The east part is planned as an exhibition area of old Beijing folk culture. Some of the featured and historical courtyards are combined with the market to develop modern industries such as culture gallery, folk-custom master studio and so on.

Meanwhile, measures are taken to protect the street more deeply.

1)To protect as original: Keep the layout, width and floor materials of streets and lanes, and and ancient building like archway and old shops keep as original to show the history.

2)To reconstruct at original sites: reconstruct historical buildings which have hidden danger at their original sites.

3)To protect historic information: protect the continuity of original functions, appearances and names of historic information to bring Beijing charm of historic culture alive.

4)To protect decorative details of buildings: include plaques, shop signs and other details to show the beauty and gorgeousness of historical building decoration.

街区规划以"保护为主、突出重点、有机整合、重在创新、精心营造、持续发展"为规划原则，力求达到历史文化、古迹遗存、传统风貌的整体性保护与重构历史风貌相统一，历史文化资源价值与激活现代功能、效益相统一。

大街两侧建筑按具体情况分为保留建筑修复风貌、历史建筑原状修缮、历史建筑原址重建、更新建筑风貌控制4类，依据毗邻的4条街和22条胡同，分为5个功能区，分别为中华老字号区、民俗餐饮区、高档商品区、四合院体验区和娱乐休闲区。

前门东片规划为老北京民俗文化的展示聚集区，发展胡同游、四合院深度体验游；对于一些有特色和历史的院落，则结合市场开发现代产业，如文化展览馆、民俗大师工作室或非遗传习所等。

同时，采取多重措施，实施深层次的保护。

1）原地原样保护。历史街巷肌理、道路位置及走向原状保护，以再现城市肌理的"文脉""气脉""地脉"。道路宽度、地面材质保持原状（包括前门大街、鲜鱼口街）；前门牌楼、古店、名店等历史建筑，原地原样保护，再现历史的"真实""真切""真情"。

2）原地复建。对存在隐患的历史建筑进行原地复建，再现历史风貌（包括广和剧场、前门牌楼、正阳桥）。

3）历史信息的延续性保护。对历史文化信息的原功能、原风貌、原街名进行延续性保护，以再现历史文化的"京味""京韵"（包括以全聚德烤鸭为代表的老店，台湾会馆等多种会馆文化，所有街巷、胡同名字及逸事传闻等）。

4）建筑细部装饰艺术展示保护。包括店牌匾、幌子、各细部装饰等，再现历史建筑装饰的"富丽""俏丽""亮丽"。

Street Design Features 街区设计特色

Color Design

In the application of color, grey is the major color supplemented with red, yellow, blue, green and white. This unique color design creates a dignified, elegant and harmonious environment which combines antique and modern charm.

Chinese red: red is a color of Beijing and also a symbol of China. Red palace walls, red lanterns, red wedding ceremony and red spring festival scrolls compose an identified color of Beijing.

Glaze yellow: it is a unique color of street scenic representing the unique splendor and glory of natural landscape, culture and history of Beijing.

Chinese Scholar tree green: Scholar tree is the city tree of Beijing. It is a symbol of life and nature which implies treasuring homeland and developing harmoniously with nature.

Blue-and-white blue: gentle and graceful, this color has a sense of historic beauty. It symbolizes civilization and creation.

Great Wall grey: the grey of courtyard residences hiding in green trees is an important symbol color of traditional buildings and landscape in this street.

Jade white: wearing jade is a symbol of moral cultivation. Mutton-fat jade is the best in jades, therefore jade white is a symbol of good luck and happiness.

Urban Furniture Design

The urban furniture design is base on Chinese traditional element and is design by modern technique to enable Chinese culture diffuse fashion breath in the street. Its color is also following the color planning system of Beijing and thus, its shape and color are perfectly combined with surroundings to add vitality to Qianmen Street.

色彩设计

在色彩的使用上，前门大街以灰色为主，辅以红、黄、蓝、绿、白等颜色作为点缀。独特的色彩设计手法营造了庄重、大气、素雅、和谐的环境，使古韵与现代相结合。

中国红。红色是北京的颜色，也是中国的象征。红色的宫墙，红色的灯笼，红色的婚礼，红色的春联……红色，构成了人们认同北京、认同街区的颜色。

琉璃黄。黄色的琉璃瓦，金秋的树叶和丰收的农田，是北京最亮丽的色彩。“琉璃黄”是街区风光特有的颜色，代表着北京独特的自然景观、人文与历史的精彩和辉煌。

国槐绿。国槐是北京市市树，郁郁葱葱的国槐是自然的风采，是生命与环境的象征。国槐绿寄寓着街区珍视自己的家园、与自然和谐发展的美好向往。

青花蓝。温润而典雅的“青花蓝”，具有一种历史的美感，是街区丰富多彩的艺术宝藏中极具代表性的色彩，象征着文明与创造。

长城灰。掩映在绿树丛中的四合院民居的灰色是街区传统建筑景观中的重要标志色。

玉脂白。“君子佩玉”——自古以来中华民族以佩玉为道德修养的标志。“羊脂玉”更是玉中极品。玉脂白是街区吉祥如意的象征。

城市家具设计

城市家具设计以中国的传统元素为基础，加以现代化的设计，使得中华文化在街区里散发时尚的气息。在色彩上遵循北京市的色彩规划体系，从而使得形色与周围环境完美融合，为古韵时尚的前门大街添加了几许灵气与活力。

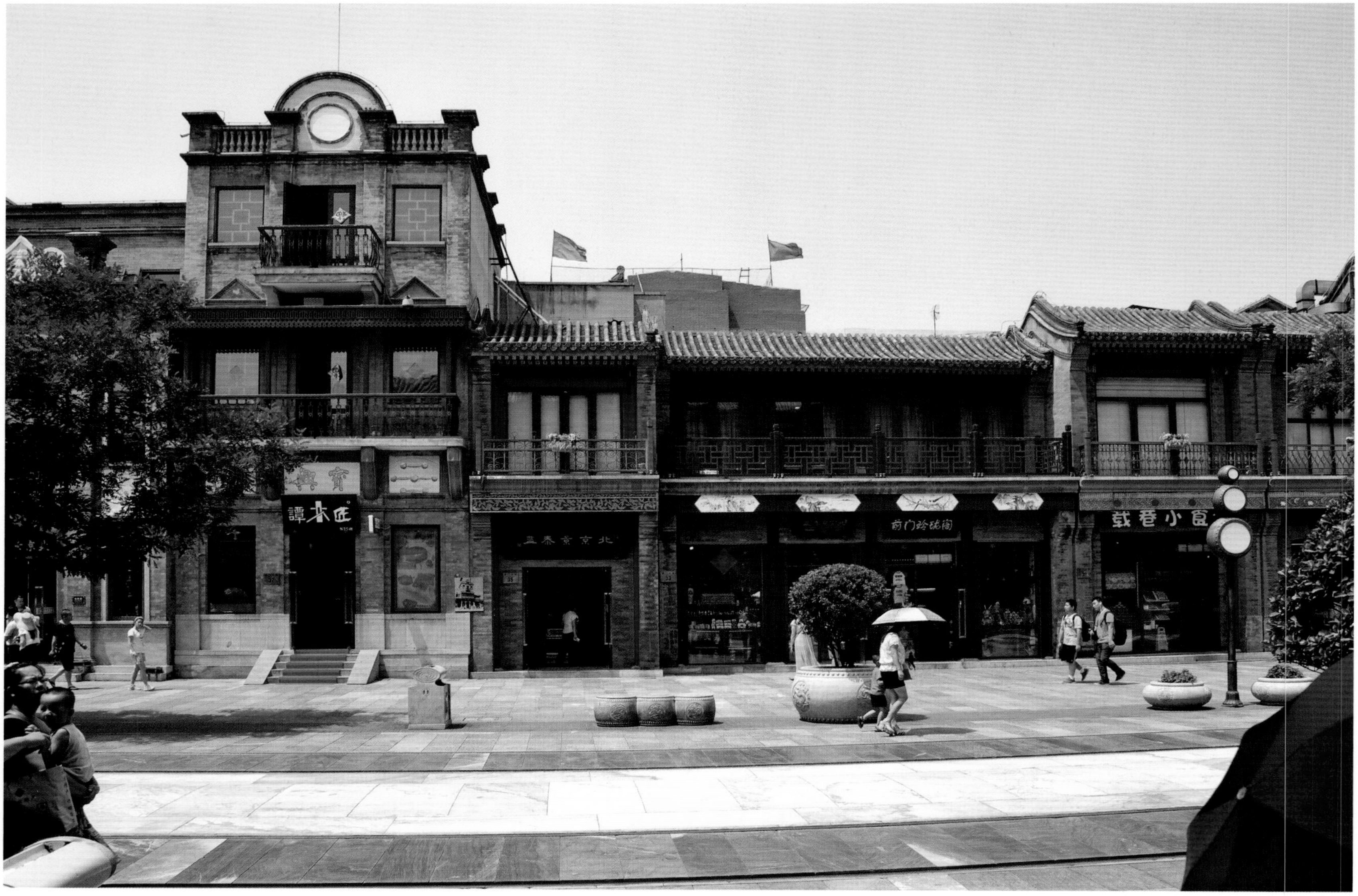

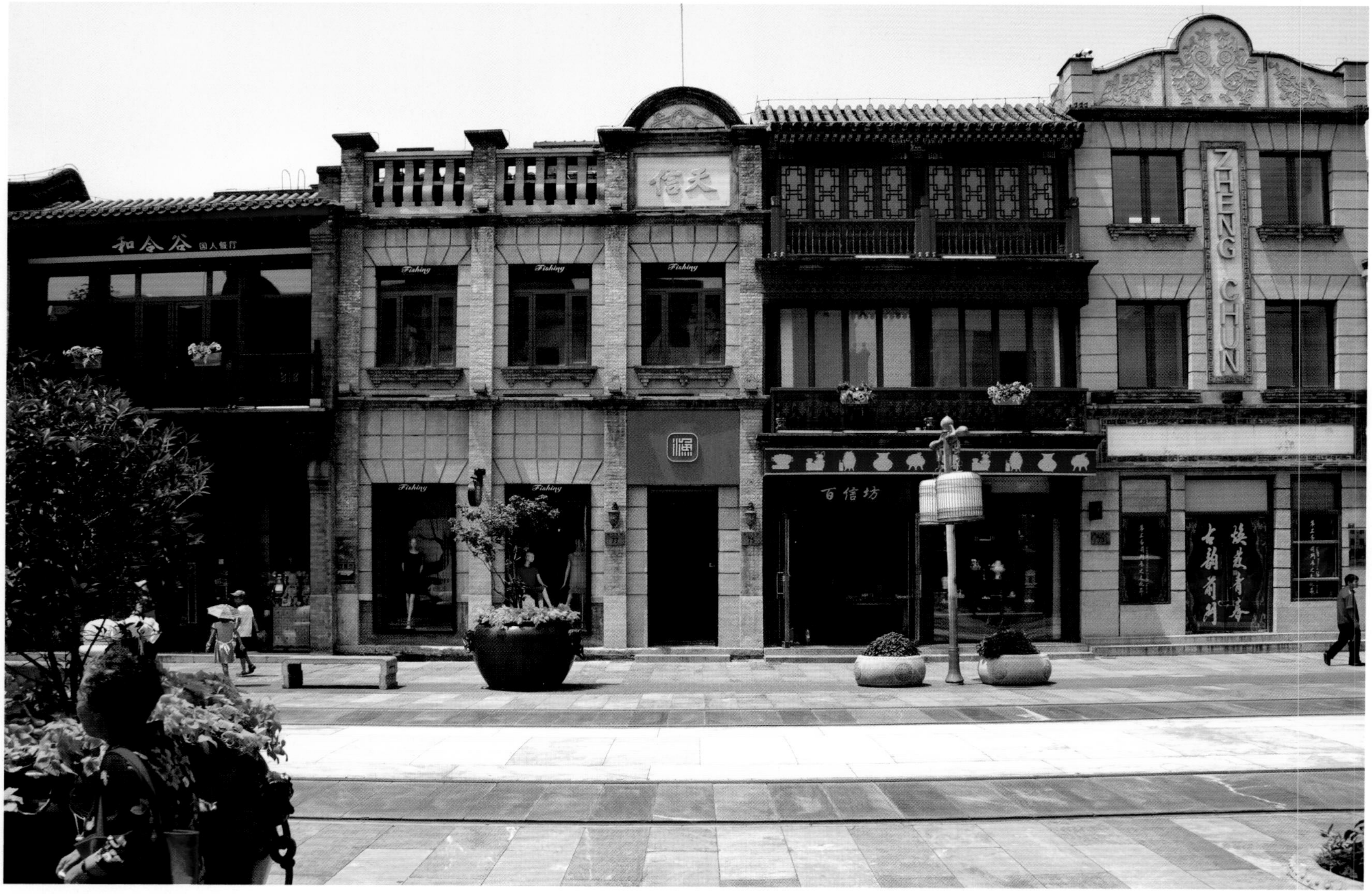
和合谷
Fishing
ZHENG CHUN
百信坊

nxilu
楼 和 廣

宣文楼
老北京
炸酱面老店
前街老铺爆肚
HUYUTAI TEA

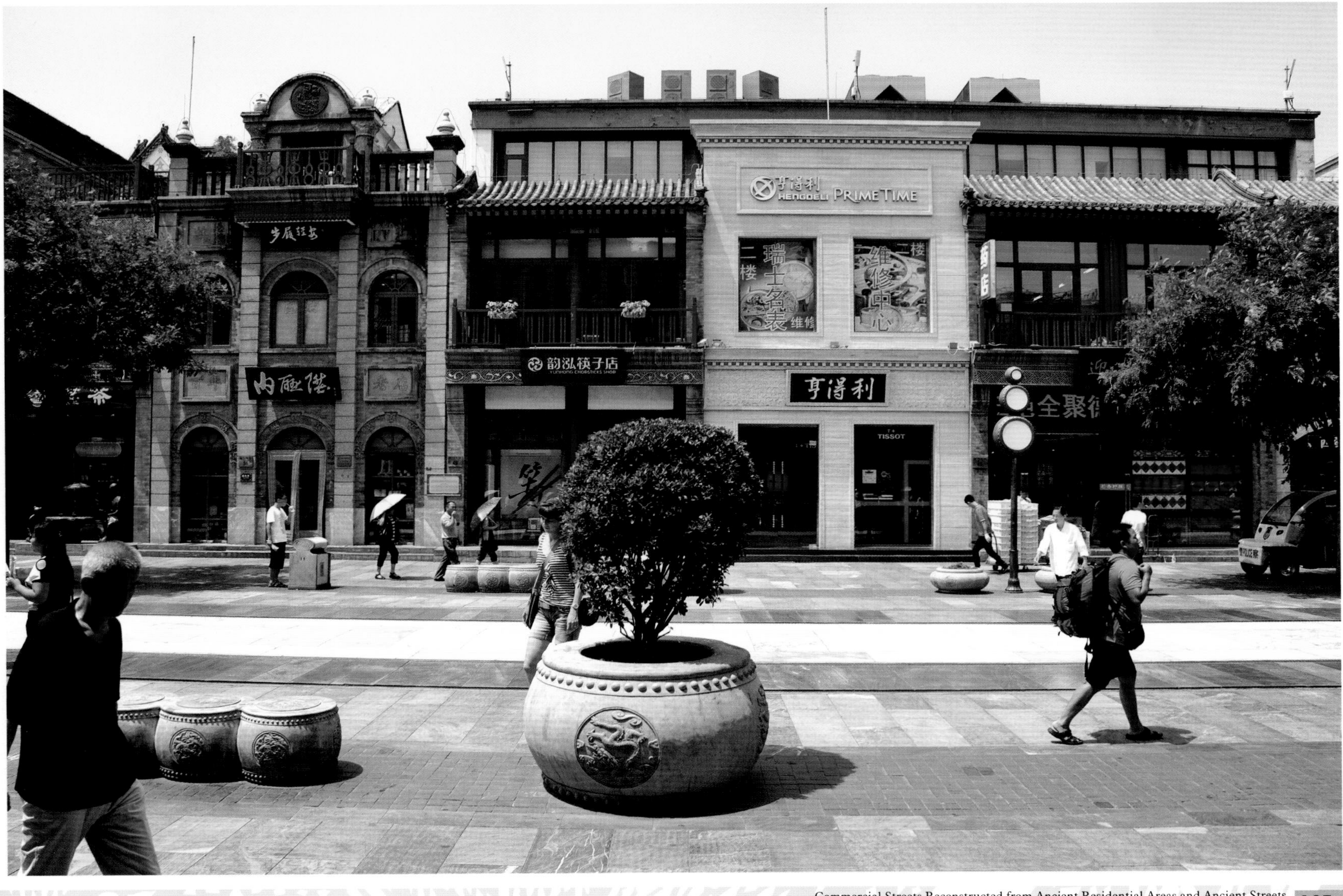

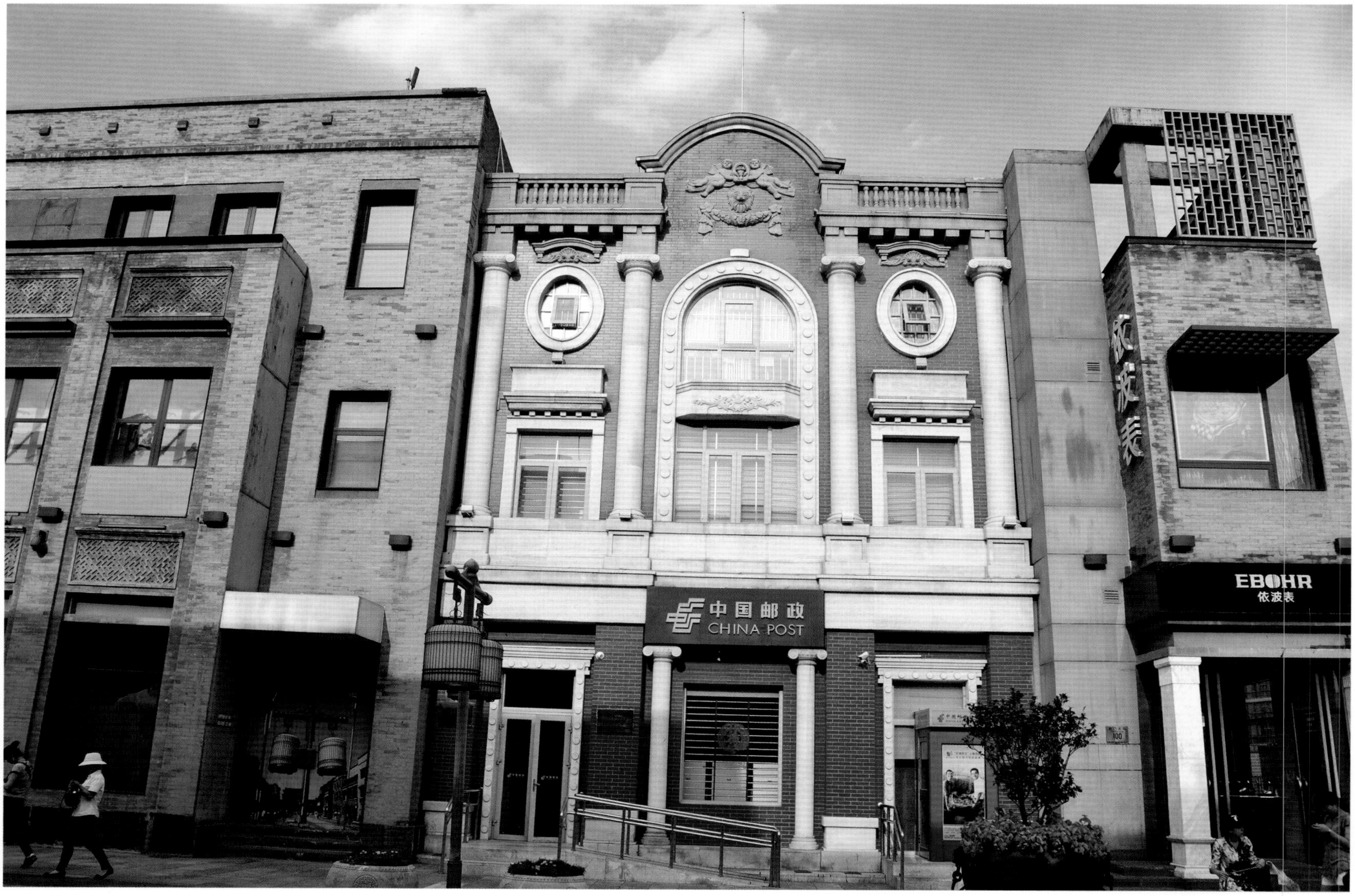
中国邮政
CHINA POST
EBOHR
依波表

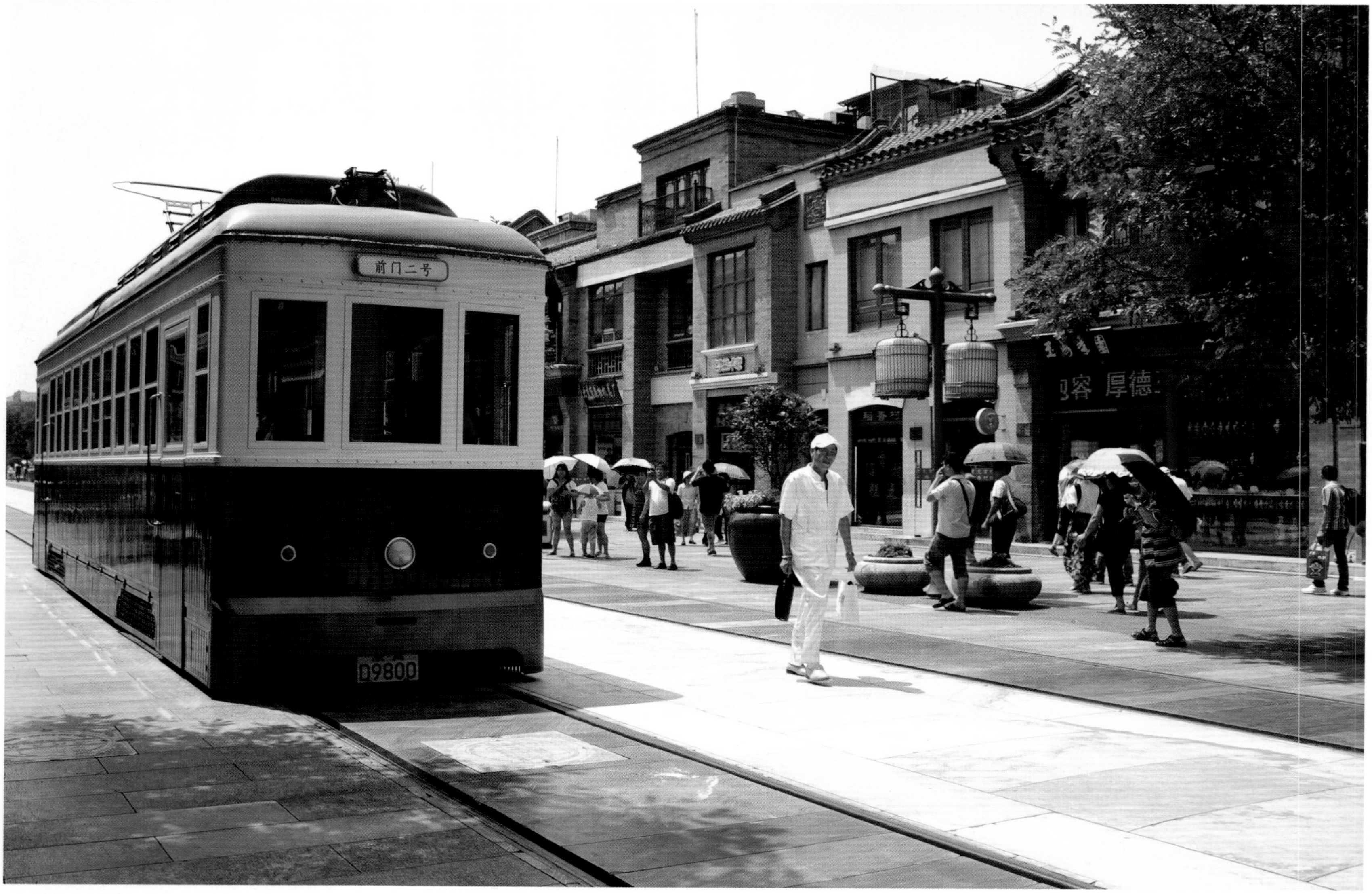
前门二号
D9800

廊房頭條胡同

中和戲院
招待所
鲑鱼 羊肉串
北京炸酱面

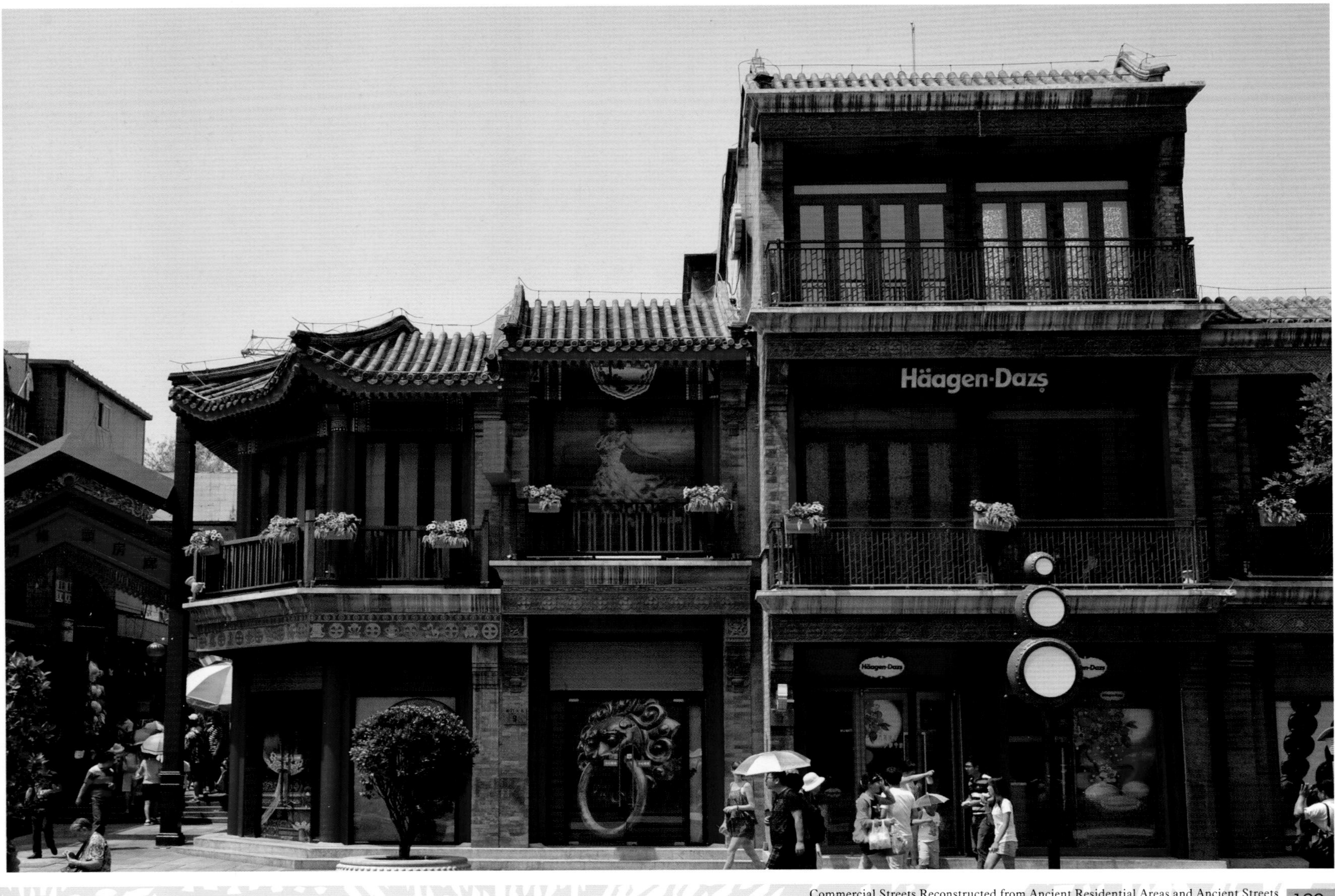

Major Commercial Activities 主要商业业态

In Qianmen Street, at the east side there is Dabei Photo Studio, Qinglinchun Tea Shop, Tongsanyi Dry Seafood Shop, Lili Restaurant, Laozhengxing Restaurant, Qianmen Hardware Store and so on; at the west side there is Huafu Watch & Clock Shop, Qingyitang Drugstore, Shengxifu Hattery, Gongxing Cultural Goods Shop and so on. After 1979, the existing old shops and traditional operating features are preserved, meanwhile new shops of hardware and electrical equipment, clothes, bicycle, food, clock, and chemical engineering are opened successively.

Qianmen Street has very long history and has brought up many old famous brands, such as Liubiju, Tongrentang, Ruifuxiang, Neiliansheng and so on. All together there are 16 old famous shops at both sides of the street. While in hutongs like Xianyukou, Damochang, there is no old famous shop, there are rough hair salons and small restaurants operated by outsiders.

前门大街东侧从北往南有大北照相馆、庆林春前门大街茶叶店、通三益果品海味店、力力餐厅、天成斋饽饽铺、便宜坊烤鸭店、老正兴饭庄、普兰德洗染店、亿兆棉织百货商店、前门五金店等店铺；西侧从北往南有月盛斋酱肉铺、华孚钟表店、庆颐堂药店、一条龙羊肉馆、盛锡福帽店、公兴文化用品店、祥聚公饽饽铺、龙顺成木器门市部、前门大街麻绳店、前门自行车商店、前门信托商店等店铺。1979年以后，在原有老字号商店和传统经营特色基本保留下来的同时，又陆续开设了五金交电、服装百货、自行车、食品、钟表、化工油漆等新店。

前门大街历史悠久，造就了许多中华老字号，如六必居酱园、同仁堂药店、瑞蚨祥绸布店、长春堂药店、内联升鞋店、张一元茶庄等。还有月盛斋酱肉店、都一处烧卖店等16处老字号分列道路两侧。在鲜鱼口、打磨厂等胡同内，店铺、饭馆林立，没有名声的老字号，多是外地人在此开设的条件简陋的美容美发店、小饭馆。

Xianyukou Food Street

Xianyukou Street extends from Qianmen Street at the west to Qianmen East Road at the east. Buildings at its two sides are mostly Republic of China style houses of two or three floors. And the lanes keep the layout of Ming and Qing dynasties.

Xianyukou was built in Ming Dynasty and gradually developed with the river ways for transport. During Zhengtong Period of Ming Dynasty, a moat was built and the mouth of the moat is dredged to drain water, therefore, Sanli River was formed. Where there is water there is fish, thus fishermen used to sell fish around the west side of a bridge where gradually became a market and called Xianyu Lane. Untill Qing Dynasty when the Sanli River dried up and shops and houses were built there, it began to be called Xianyukou.

鲜鱼口美食街

鲜鱼口街西起前门大街，与大栅栏隔街相连，东至前门东路，与西兴隆街相接，是一条占地1.3公顷、全长225米的东西向商街。街巷两旁多为二三层的民国风格小楼，街巷走向保持着明清时期的布局。

鲜鱼口始建于明代，随着高粱河旧道及漕运而逐步发展起来。明正统年间，修城壕，为了方便夏季雨水排出，故于正阳桥东南低洼处开通壕口，以泄其水，形成了三里河。据《日下旧闻考》记载，“正阳门外东片有三里河一道。”因为有了水，就有了鱼，也有了渔民在此打鱼。渔民打到鱼后经常拿到鲜鱼口内小桥西侧售卖，渐成集市，人称鲜鱼巷。到了清代，三里河的水道干涸，其上慢慢搭建了店铺和房屋，始称鲜鱼口。1965年，小桥等三条胡同并入，统称鲜鱼口街，向东延伸至现在的位置。

Liulaogen Stage

Liulaogen Stage Theatre was founded by Zhao Benshan in Shenyang and has become a chain theatre. It is so popular that its tickets are always sold out. And now it is a tourism name card of Shenyang.

Liulaogen Stage in Beijing was opened in May 3rd, 2009. It began to officially sell tickets after trial performance. Halls of the stage are named after TV plays and dramas produced by Zhao Benshan.

刘老根大舞台

2003年4月，辽宁民间艺术团成立。不久，赵本山决定在沈阳创建“刘老根大舞台”剧场。到2007年4月，“刘老根大舞台”剧场已增加到6家。2008年春天，位于沈阳中街的“刘老根大舞台”翻新建成大剧场。2009年，其在沈阳、长春、天津、哈尔滨、北京等城市开办了8家连锁剧场，一年四季一票难求，现已成为沈阳的一张旅游名片。

刘老根大舞台北京剧场于2009年5月3日正式开张，经过试演后正式对外售票演出。大舞台的大厅分别以赵本山出品的电视剧和小品命名。

刘老根大舞台
过大年
赶大集

Dajiang Hutong

Dajiang Hutong appeared in Ming Dynasty when it was called Jiangjia Hutong. It was changed its name into Dajiangjia Hutong in 1965. It is over 600 meters long and is a shortcut from Qianmen Street to Zhushikou East Street. There are famous Guild Theatrical Stage, Guozi Market, Cloth Lane, and Embroider Street which are vivid reflection of lives in old Beijing.

大江胡同

大江胡同在明代就已出现，称蒋家胡同，清代改称大蒋家胡同，1965年改称为大江胡同。大江胡同有600多米长，是从前门大街通往珠市口东大街的捷径。在大江胡同包围的扇形地段里，有京城著名的会馆戏楼、果子市、布巷子、绣花街、老冰窖，这些都是老北京生活的生动写照。

Ali Mountain Square

There are many shops at Ali Mountain Square. Gelaifu Trade Company at No.126 Dajiang Hutong on Qianmen Street mainly sales nougats, pineapple cakes, chocolates and so on, which can not only be enjoyed by oneself, but also be used as gifts in the festivals.

阿里山广场

在阿里山广场有很多商铺，位于前门大街大江胡同126号的阁来福商贸有限公司主要出售台湾義美食品，如牛轧糖、凤梨酥、巧克力、竹炭花生等，既可以买来自己享用，也适合在节日期间馈赠亲朋。

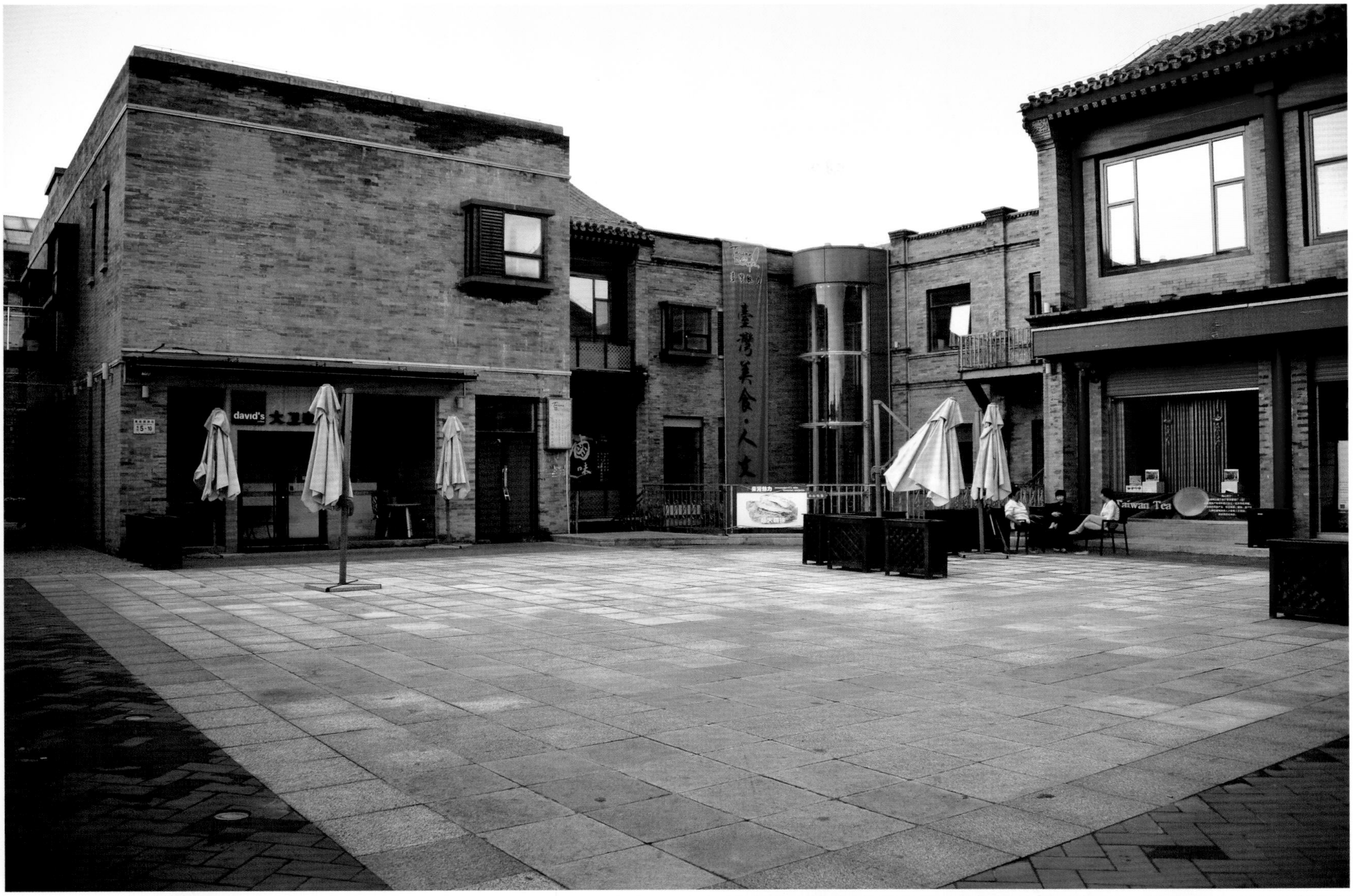

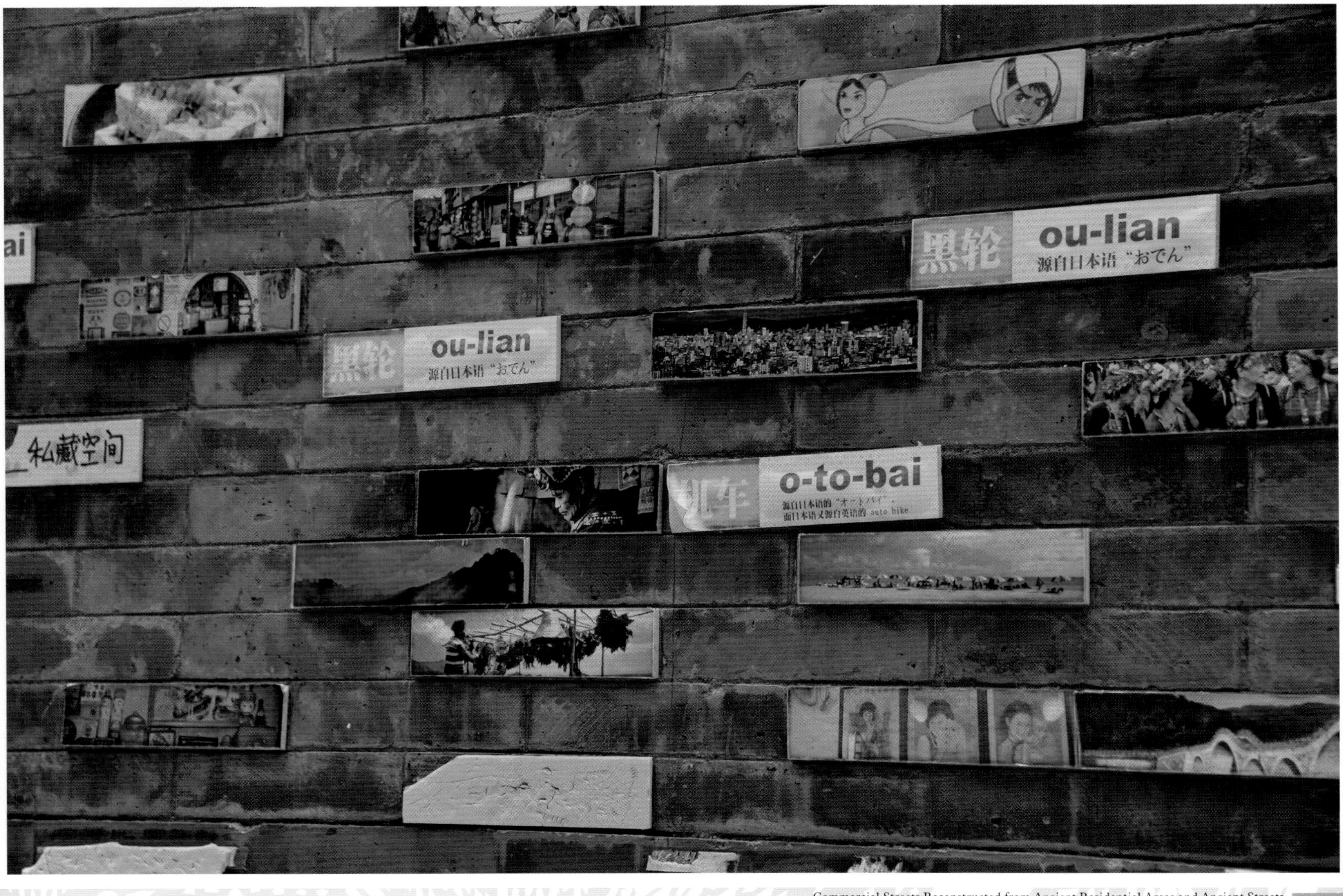

Taiwan Snack Street

Taiwan Snack Street has almost all kinds of snacks of southern Fujian. This small street gathers delicious snacks from the three most well-known night markets — Taipei Shilin Tourist Night Market, Taichung Fengjia Night Market and Kaohsiung Liuhe Night Market. There is a story after every snack. Visitors can satisfy their appetite to the full.

台湾小吃街

台湾小吃街有最全的闽南小吃。小小的街巷荟萃了台湾知名度最高的台北士林观光夜市、台中逢甲夜市、高雄六合夜市等三大夜市的美味小吃。大鱼丸汤、牛肉丸汤、度小月担仔面、月亮虾饼、彰化肉圆，每种小吃的背后都有一段小故事。蚵仔煎、生炒花枝、海鲜粥、鱿鱼羹、虱目鱼汤等，滋味鲜美，往往让人吃得大呼过瘾。

Food & Beverage 餐饮类

Quanjude

A time honored restaurant, founded in 1864. It is the first "China Famous Brand" in service industry. After ceaseless innovation and development, its dishes featuring roast duck and gathering more than 400 dishes are favored by heads of states, officials, people of various fields and visitors from home and abroad. Its dishes are praised as "the top cuisine of China".

全聚德

全聚德为中华著名老字号，创建于1864年（清朝同治三年），历经几代的辛勤拼搏获得了长足发展。1999年，全聚德被国家工商总局认定为“驰名商标”，是我国第一例服务类中国驰名商标。菜品经过不断创新发展，形成了以独具特色的全聚德烤鸭为龙头，集“全鸭席”和400多道特色菜品于一体的全聚德菜系，备受各国元首、政府官员、社会各界人士及国内外游客喜爱，被誉为“中华第一吃”。

全聚德挂炉
National Intangible Cultural Heritage
Navig ation

Tianfu Hao

"Tianfu Hao Pork Shoulder" is a famous food in Beijing. Its fame has been lasting over 200 years, for the braised pork shoulder is of good looking, fragrant and tasteful.

天福号

"天福号酱肘子"是北京特色风味名食，从200年前创业至今盛名不减，令人称奇。酱肘子肉皮酱紫油亮，鲜香四溢，入口无油腻之感，回味长久，美食扬名后世。

Hou Wonton

There are five favors of wonton and about one hundred dishes. Now the cooking of wonton is different from before which boiled ed all wontons in a big cauldron. Now wontons are cooked in a small basket. And the soup is cooked previously and is divided into several kinds which taste fresh and delicious.

侯馄饨

侯馄饨店内光馄饨就有5种风味，还有南北小吃，炒菜达百种之多。以前煮馄饨的方法是将一簸箕包好的馄饨往大锅一倒，煮好后，连汤带馄饨往碗里一盛，就算齐活。现在改用小篓(一篓10个)单煮，汤用棒骨、老鸡夜间吊好，在吊锅里熬，且分菜肉、虾肉、鲜肉、红油和酸汤若干种，口味鲜香，而且质量稳定。

Milktea

Milktea is a small specific store mainly sells milky tea. Its decoration is novel and fashionable, and its furnishing is simple, elegant and romantic. It is a good choice for white collars, business travelers and lovers.

Milktea

Milktea是一家以奶茶为主的特色小店，以新颖时尚的装修特色展现于众，店内布置简约、高雅，弥漫着温馨与闲情，是精英白领、商务旅客、情侣约会的时尚之选。

Starbucks

Founded in 1971, now Starbucks is the biggest coffee chain with its headquarters locating at Seattle. Apart from coffee, Starbucks also provides tea, pies and cakes. It has about 12,000 branches all over the world, including North America, South America, the Middle East and the Pacific region.

星巴克

星巴克于1971年成立，为全球最大的咖啡连锁店，总部坐落于美国华盛顿州西雅图市。除咖啡外，星巴克亦有茶、馅饼、蛋糕等商品。在全球范围内已经有近12 000间分店，遍布北美、南美、欧洲、中东及太平洋地区。

Others 其他

Qianxiangyi

Qianxiangyi is an old silks and satins shop of almost 80 years old. It is known for its complete goods of various demands and considerate services. With operating ways of fixed prices, buy ten get one, home delivery service and so on, it has a strong competitiveness in the market.

谦祥益

谦祥益是一家有近80年历史的老字号绸缎店，是山东章丘旧军孟氏进修堂创办的"祥"字号之一，以货色齐全、服务周到而闻名。在经营中，采取敞架售货、明码标价、加一放尺（买10尺送1尺）、送货上门等方式，营业额与日俱增，在激烈的竞争中长盛不衰。

Xiefuchun

Xiefuchun has a history of 177 years, and is one of ancestors of China's cosmetic industry. It was founded in 1830 by Xie Hongye. The name of Xiefuchun has a meaning of forever youth in China.

谢馥春

谢馥春已有177年的历史传承，是中国化妆品业的始祖之一，创建于清朝道光十年（1830年），创始人谢宏业取“谢馥春”为店名，“谢”为姓，汉语中有凋零衰败之意，故加“馥春”二字，“馥”意为馥郁芬芳，并与“复”谐音，与“春”字相连，寓意青春永驻。

Yizhao Department Store

Yizhao means booming business and being wealthy. On the base of preserving traditional products, it adds some new wool products of modern technology according to the demands of market. At present, Yizhao Department Store provides over 150 kinds of wool products of more than 100 colors.

亿兆百货

亿兆之名，取生意兴隆、买卖发财、成亿万富翁之意。在保留原有毛线、毛衣、毛织用品等传统商品的基础上，针对目前消费者选择毛线由原来的纯毛、保暖性能好向可机洗、防缩水、防静电等需求的转变，商场增加了具有科技含量的新品毛线。目前，“亿兆”有羊绒毛线、羊毛线、棉毛线、丝光线、棒针毛线、意大利进口毛线、日本毛线等针棉新品货源150余个品种，近100多种颜色。

Chow Tai Fook

Chow Tai Fook mainly retails, wholesales and produces jewelries. It has a wide range of products whose design is fashionable, unique and tasteful. Meanwhile, Chow Tai Fook is one of the worldwide 78 sightholders of DTC, the biggest raw diamond supplier in the world.

周大福

周大福主要经营珠宝首饰零售、批发和制造业务。货品种类繁多，款式新颖独特，极具品位。同时，周大福是全球最大钻石毛胚供货商英国钻石贸易公司全世界78个特约配售商之一，获直接配售天然钻石原胚，是整个大中华地区首家获得此殊荣的钻石商，地位超然。

Wu Yutai

Wu Yutai teahouse was founded in 1887 by Wu Xiqing. It mainly provides production, process, packing, delivery, wholesale and retail of tea-leaf, tea products, and tea derivatives.

吴裕泰

吴裕泰茶庄原名吴裕泰茶栈，始建于清朝光绪十三年（1887年），创办人是吴锡清，当时是为吴氏家族茶庄进储茶叶而建的。吴裕泰主营茶叶、茶制品、茶衍生品的生产、加工、拼配、分装、配送、批发与零售等。

Samsung

Samsung Group is the biggest enterprise in South Korea including 26 subsidiary companies and several other corporations. Its business involves fields of electron, finance, machinery, chemistry and so on.

三星

三星集团是韩国最大的企业集团之一，包括26个下属公司及若干其他法人机构，在70个国家和地区建立了近300个办事处，员工总数196 000人，业务涉及电子、金融、机械、化学等众多领域。

ZARA

ZARA is a subsidiary corporation of Inditex Group. Inditex is a clothing retailer ranking first in Spain and ranking third in the world with eight clothing retailer brand among which ZARA is the most famous one.

ZARA

ZARA是西班牙Inditex集团旗下的一个子公司，Inditex是西班牙排名第一、全球排名第三的服装零售商，在全球64个国家和地区开设了2 899家专卖店，旗下共有8个服装零售品牌，ZARA是其中最有名的品牌。

UNIQLO

UNIQLO is a famous Japanese clothe brand founded by Tadashi Yanai in 1984. It offers fashionable and high quality casual wears of novel design to consumers of varied age groups.

UNIQLO

UNIQLO是日本著名的服装品牌，由日本人柳井正创立于1984年。向各个年龄层的消费者提供时尚、优质的休闲服，服装款式新颖，质地细腻。迄今为止在6个国家拥有760家分店、20 000名雇员。

New Balance

It was founded in 1906 by William J. Riley in Boston. Now it has become a favorite brand of many successful entrepreneurs and political leaders. It is praised as "president jogging shoes" in America and many other countries.

新百伦

该品牌也译作“纽巴伦”，1906年，William J. Riley在美国马拉松之城波士顿成立，现已成为众多成功企业家和政治领袖爱用的品牌，在美国及许多国家被誉为“总统慢跑鞋”“慢跑鞋之王”。

A-Panda

It is a specific shop engaging in developing, designing, makeing and selling tourism products.

阿盘达

阿盘达是一家从事旅游产品开发、设计、制作、销售的特色店铺。

Sephora

Sephora was founded in 1969 in Limoges and joined in the global top luxury brand LVMH in 1997. Now it has 1 000 shops all over the world, selling skin care products, cosmetic products and perfume.

丝芙兰

丝芙兰，1969年创立于法国里摩日，1997年加入全球第一奢侈品牌公司LVMH。目前在全球拥有1 000家门店，店内经营护肤品、美容产品和香水等。

引导指示系统

Guidance & Sign System

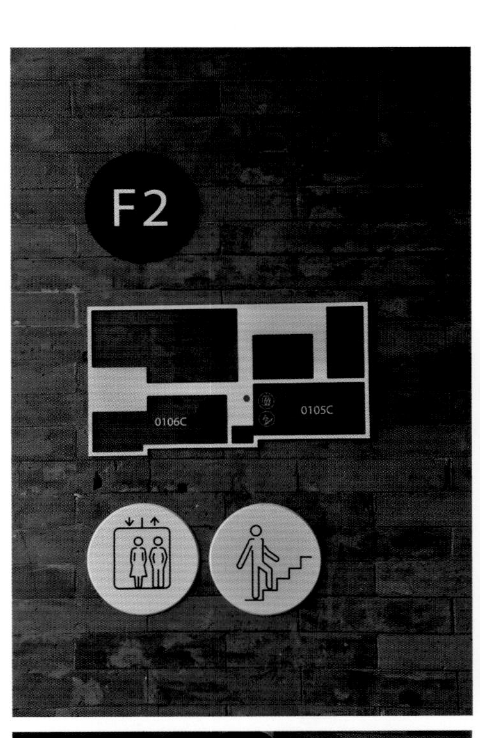

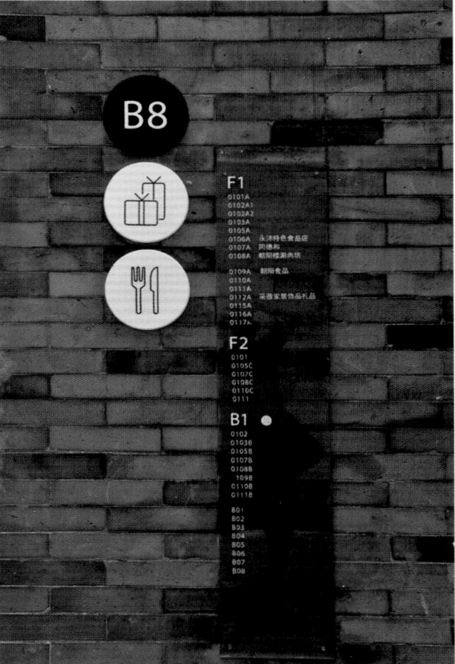

Dashilan, Beijing
北京大栅栏

街区背景与定位
Street Background & Market Positioning

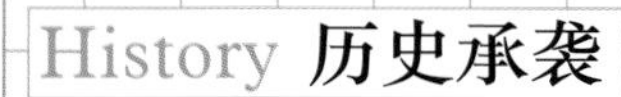

History 历史承袭

Springing up in Yuan Dynasty, being founded in Ming Dynasty and thriving in Qing Dynasty, Beijing Dashilan has a history of 500 years. In 1488, the emperor gave orders to set up barriers in Dajiequxiang and sent soldiers to prevent steals. Until Qing Dynasty, Dashilan developed into a primary commercial center. Actually, this street was originally called Langfangsitiao but changed its name to Dashilan which means big barriers because of the barriers of outstanding quality.

In 1900, the Dashilan was burnt down by the Boxer. The rebuilt Dashilan is as prosperous as usual. At present, Dashilan has recovered the style and features of early Republic of China.

北京大栅栏（读作“大石烂儿”）兴起于元朝，建立于明朝，从清朝开始繁盛至今，已有500余年的历史。早在明朝弘治元年（1488年）明孝宗下令在北京城内大街曲巷设立栅栏，并派兵把守以绝盗窃。到了清朝，大栅栏发展成为主要的商业中心，据《钦定令典事例》记载，雍正七年批准的外城栅栏440座，乾隆十八年批准的内城栅栏1 919座，皇城内栏196座。其实，大栅栏早先并不叫大栅栏，而叫廊坊四条，后来就因为这条胡同的栅栏制作出色，保留长久，而且偏大，为京城所瞩目，因而大栅栏就成了这条胡同的名称。老北京有句顺口溜：“看玩意上天桥，买东西到大栅栏。”

1900年，义和团曾将大栅栏整条街焚毁，重建大栅栏后，街区繁华依旧。现今的大栅栏复原了民国初期的风貌。

Location 区位特征

Located at the center section of Beijing - at the south of Tiananmen Square and the west of Qianmen Street, Dashilan is totally 275 meters long and is a important part of the axis of the south Beijing. Dashilan has a long history and profound culture. It has a convenient traffic and attracts a large number of tourists; therefore it enjoys exceptional advantages in business and location.

大栅栏位于北京市的中心地段，是北京城南中轴线的一个重要组成部分，在天安门广场的南边，前门大街的西边，从大栅栏的东口至西口全长275米。大栅栏历史悠久，文化源远流长。地段的东、南面为商业型城市干道，交通便利，游客众多，具有得天独厚的商业、旅游区位优势。

Market Positioning 市场定位

Committing itself to developing cultural business, cultural tourism and relative exhibitions, Dashilan aimed to be a commercial culture base that assembles business, entertainment and tourism.

大栅栏致力于文化商业、文化旅游业及相关文商博览、文商表演业的发展，凸显文商品牌，将大栅栏打造成集商务、娱乐休闲、文史旅游为一体的商业文化基地。

Street Planning 街区规划

In 2004, Xuanwu District took reasonably protecting historic street, stimulating energy of traditional business and improving quality of environment as aims in the planning of reconstructing Dashilan. The function positioning of planning section is a comprehensive traditional historical protection area of traditional business, cultural tourism and residence.

According to the function positioning, the planning has put emphasis on protecting and restoring traditional style and feature of the street and reviving traditional business. As for traffic planning, Dashilan has been zoned as pedestrian street. The size of the street is kept as original with part of lanes widened.

在2004年宣武区人民政府对大栅栏地区的规划中，以“科学保护历史街区、局部合理有机更新、激发传统商业活力、提高空间环境品质”为规划目标，将规划地段的主要功能定位为：传统商业、民俗文化旅游及居住相结合的综合性传统历史保护区域。

大栅栏街区的功能定位为传统商业。因而在规划大栅栏街区时以保护街区传统风貌与肌理为主，整治和修复破败的旧有建筑，还原历史风貌，复兴传统商业。在交通规划上，将大栅栏划为步行街区，不允许机动车辆通行，减少人车干扰。街巷的宽度、走向都保持历史原貌，对胡同局部予以拓宽，具体的区域道路规划如下：大栅栏步行街，宽6.5~15米；煤市街为双向机动车道，宽25米；粮食店街和珠宝市街为单向机动车道，宽10米。

Street Design Features 街区设计特色

With a history of more than 500 years, visitors can see old shop of hundred years everywhere. All the old buildings are of traditional Chinese style, while the newly-built buildings are of style of early Republic of China.

大栅栏拥有500多年的历史，在这条步行街上百年老店随处可见，红窗灰瓦，错落有致，体现着传统的中式风格；新建筑则统一为朱红窗格牌楼、青砖灰瓦白线墙装点的仿古式建筑，并粉刷装修立面、规范牌匾，具有民国初期的风貌。

海友客栈
汉庭海友

三希堂

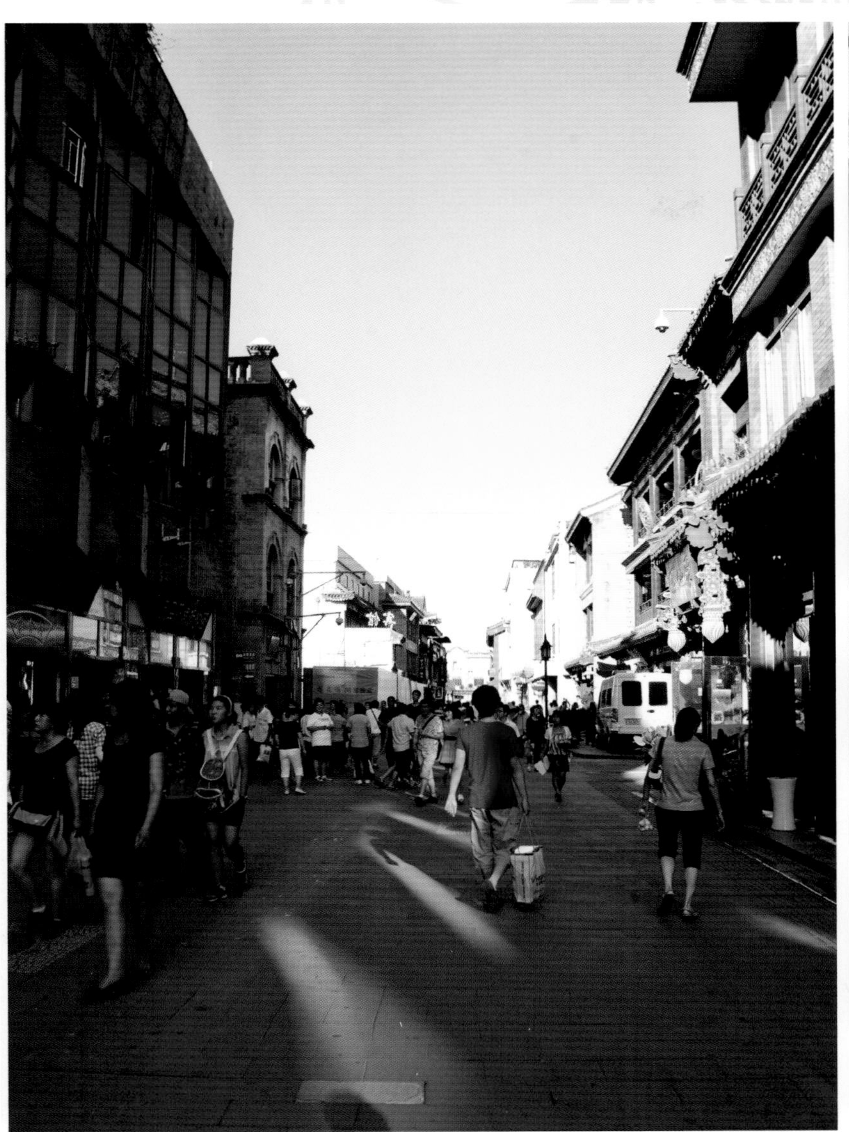

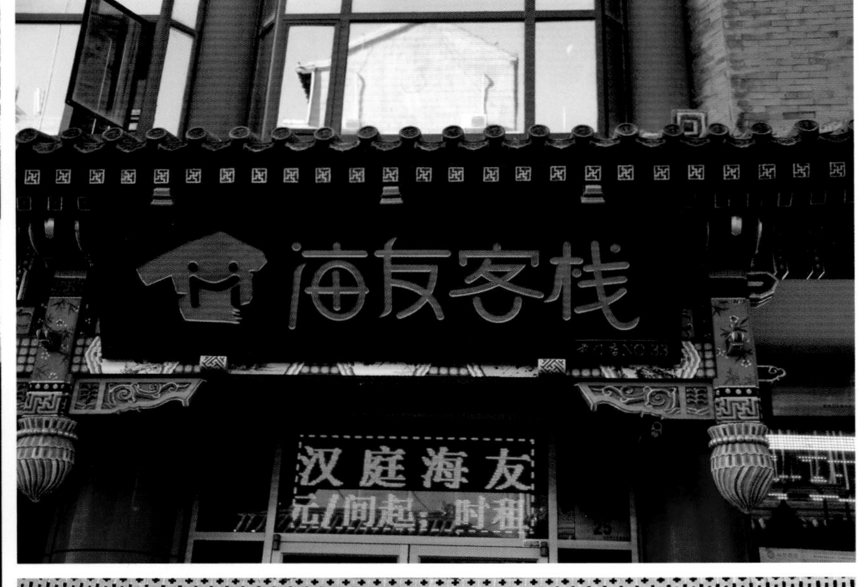

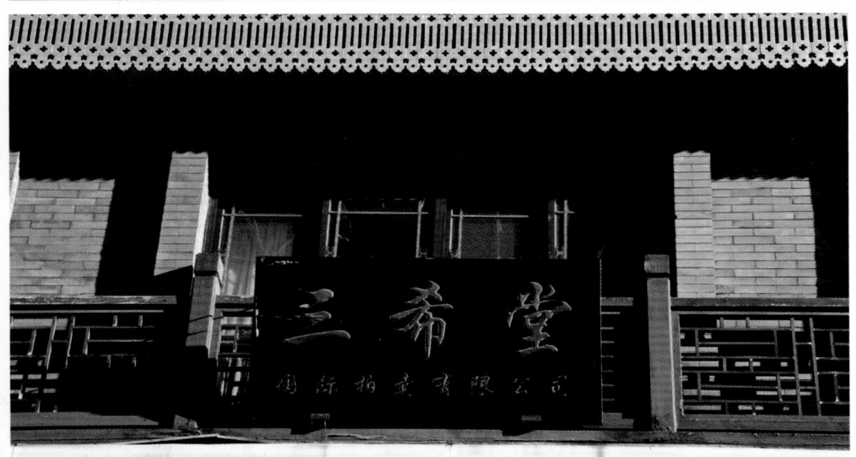

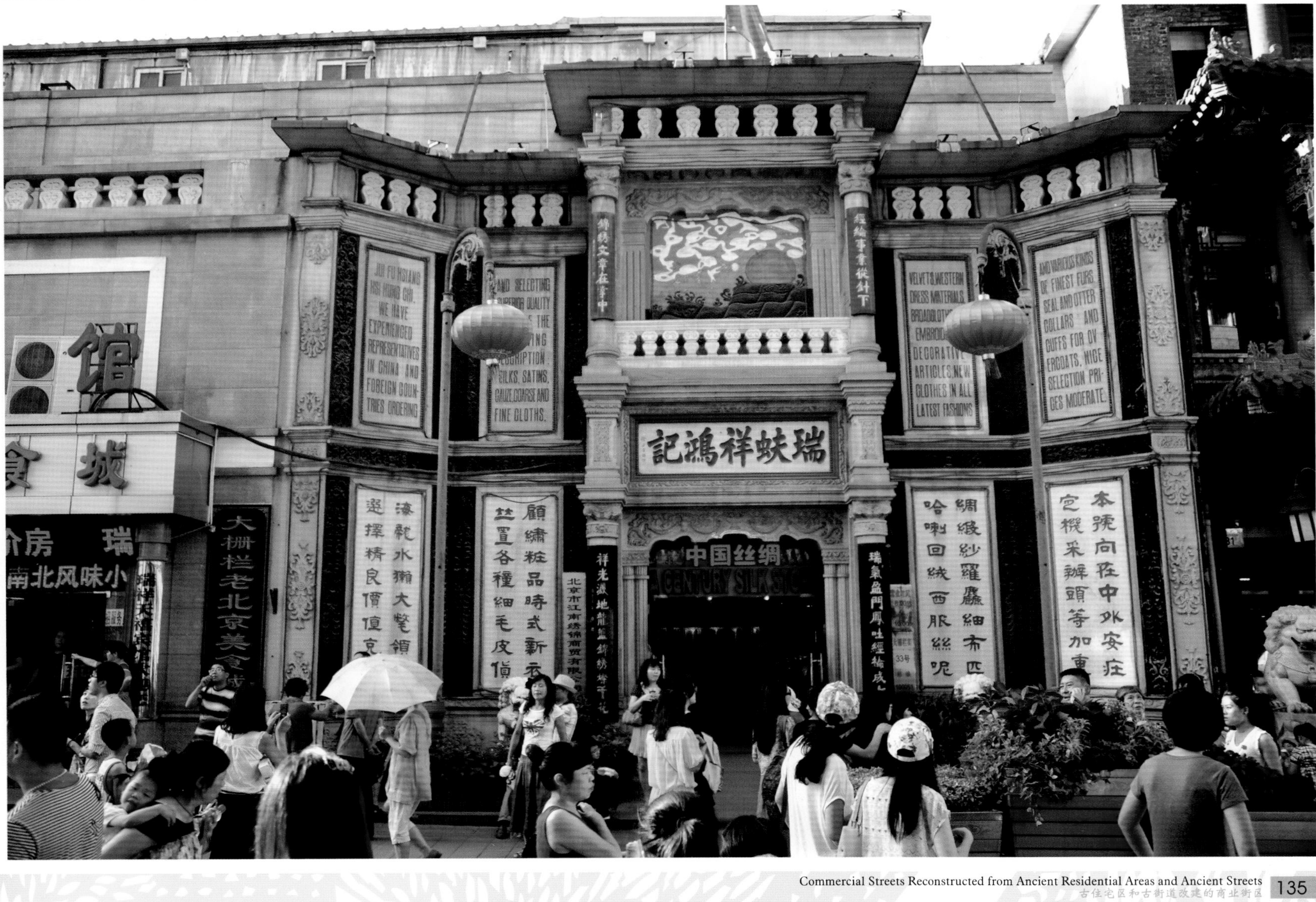
JUI FU SHIANG HSI HONG CHI. WE HAVE EXPERIENCED REPRESENTATIVES IN CHINA AND FOREIGN COUNTRIES ORDERING
AND VARIOUS KINDS OF FINEST FURS, SEAL AND OTTER COLLARS AND CUFFS FOR OVERCOATS, NICE SELECTION PRICES MODERATE.
瑞蚨祥鴻記
中国丝绸

Major Commercial Activities 主要商业业态

There are dozens of shops of 11 different industries in Dashilan. They are mainly old and famous shops. Commercial activities in Dashilan tightly relate to daily life of residents such as old shops of food and beverage, clothes, and daily necessaries. There are also old shops of culture and entertainment, and newly introduced jewelry shops.

大栅栏街区分布着11个行业的几十家店铺。店铺以老字号为主，其中的同仁堂、张一元、内联升、瑞蚨祥、步瀛斋、大观楼、狗不理、张小泉8家店就占了营业面积的1/4，年销售额约占整条街总营业收入的3/4。大栅栏的商业形态与居民的日常生活结合紧密，诸如饮食、穿着、日用品的老字号较多，另外也存在一些文化娱乐的老字号、新引进的珠宝商铺等。

Featured Commercial Area 特色商贸区

Dashilan Folk Culture Town

Dashilan Folk Culture Town is very bustling with famous Donglaishun Restaurant and busy Jilong Commercial Street. There are so many shops with numerous goods which are of national features.

大栅栏民俗文化城

大栅栏民俗文化城内有鼎鼎大名的东来顺饭庄，又有人声鼎沸的吉龙商业街，热闹非凡。街内商铺林立，商品琳琅满目，大多具有民族特色，民俗文化浓厚。

箱包 厂家直销
大甩卖

北京
泥人張
濤杰
第二届非遗文化展
泥巴狂人

Food & Beverage 餐饮类

Donglaishun Restaurant

Donglaishun Restaurant was founded by a Hui people named Ding Deshan. It was begun from a stall that sold mutton noodles in 1903. In 1914, it was renamed "Donglaishun Mutton Restaurant". And it became famous for its exquisite skill of cutting mutton later. In 1996, the headquarters of Donglaishun Restaurant Chain was founded.

Donglaishun has a history of more than 100 years for now. It is not only the model of successful combination of old famous brand and modern operation, but also a treasure of cooking culture of Muslim.

东来顺饭庄

东来顺是一位名叫丁德山的回民创立的。1903年，他在东安市场里摆摊卖羊肉杂面和荞麦面切糕，后来又增添了贴饼子和粥等食品。由于生意日渐兴隆，他挂起东来顺粥摊的招牌，寓意“来自京东，一切顺利”。1914年粥摊增添了爆、烤、涮羊肉和炒菜，同时更名为“东来顺羊肉馆”。此后，他想方设法挖来前门外正阳楼饭庄的一位名厨帮工传艺，使东来顺的羊肉刀工精湛，切出后铺在青花瓷盘里，盘上的花纹透过肉片隐约可见。 到20世纪三四十年代，东来顺的涮羊肉已驰名京城。1996年，东来顺连锁总部成立。

东来顺到现在已经有百年历史，它既是老字号品牌和现代经营结合的典范，又是清真饮食文化的一块瑰宝。

東来順

Go Believe (Gou Buli)

Go Believe was founded in 1858 by Gao Guiyou whose child's pet name is Gouzi. Gouzi opened his own shop selling steamed stuffed bun named "The Deju" three years after he studied skills in Tianjin. His business was so good that he had no spare time to talk to customers, therefore, as time passed, people called his buns as "Gou Buli bun" which means Gouzi took no notice of customers when selling bun.

"Gou Buli bun" had even got a very high praise from Empress Dowager Cixi and since then it had became very famous through out the country.

狗不理

“狗不理”始创于1858年，创始人高贵友。高贵友乳名“狗子”（根据北方习俗，名字越低贱小孩越容易养活），十四岁去天津学艺，三年期满后，狗子独立开办专营包子的店铺“德聚号”，因为手艺好，很快出名，又因为生意兴隆，卖包子时狗子也顾不得和顾客说话，人们戏称他“狗子卖包子，不理人”，久而久之，人们都叫他“狗不理”，把他做的包子称为“狗不理包子”，原店铺名却被人们淡忘。

袁世凯曾将狗不理包子献给慈禧，获得慈禧“山中走兽云中雁，陆地牛羊海底鲜，不及狗不理香矣，食之长寿也”的评论，狗不理从此闻名四海。

Zhengxingde Tea House

Zhengxingde Tea House was founded in 1738 and its founder is one of the famous "Eight Authorities" Mu Wenying, a Hui people. Zhengxingde was a small tea shop which distributed tea from Hunan, Hubei and Anhui at the beginning. After it developed its own characteristic flower tea, it became famous in Tianjin and its business was better and better.

正兴德

正兴德茶庄创办于1738年，创始人是天津著名“八大家”之一回族巨商穆文英。正兴德开设之初名为“正兴茶铺”，只是一间经销湖南、湖北绿茶及安徽大叶茶的小茶叶铺，后来研制出有自己特色的花茶，在天津走红，生意越来越好，渐渐发展壮大。清咸丰七年（1857年）改名为“正兴德记茶叶铺”。

Clothes 服饰类

Neiliansheng

Neiliansheng was founded in 1853 by Zhao Ting. Zhao Ting established Neiliansheng shoe shop rapidly after he learnt a good craftsmanship and accumulated certain customers and managing experience in Beijing.

At the beginning, Neiliansheng made use of interpersonal connections and specialized in producing court boots for aristocrats and officials. The shoes produced by Neiliansheng had become a symbol of status.

内联升

内联升始创于清咸丰三年（1853年），创始人是武清县赵廷。赵廷早年于京城学习做鞋，习得好手艺。在他积累了客户人脉和一定的管理经验后，决定自立门户，加上“丁大将军”的资助，赵廷迅速创办起内联升鞋店。

内联升创办之初就利用人脉关系，专门为皇亲国戚、文武百官制作朝靴。“内”指大内宫廷，“联升”示意穿上此店制作的朝靴，可以在宫廷官运亨通，连升三级。当时在老北京有这么一句口头禅：“头顶马聚源，脚踩内联升，身穿八大祥，腰缠四大恒。”内联升做的鞋已然成为一种身份的象征。

Ruifuxiang

Ruifuxiang, founded in 1893, is famous throughout the country and is the top of "Eight Xiangs" in old Beijing.

Profound culture deposits of over one hundred years history, unique and substantial operation and management and management aim of "sincerity, honesty, no bargaining and no cheating" make Ruifuxiang exist for over hundred years and always keep it leading position in silk industry and hand-tailor industry.

瑞蚨祥

瑞蚨祥始创于清光绪十九年（1893年），是海内闻名的中华老字号、旧京城“八大祥”之首。北京城曾流传着这样的歌谣：“头顶马聚源，身穿瑞蚨祥，脚踩内联升”，这是对瑞蚨祥当时繁盛的生动描写。

百余年的发展使瑞蚨祥有着深厚的文化底蕴，加上它独特而颇具实力的经营管理手段，并贯彻“至诚至上，货真价实，言不二价，童叟无欺”的经营宗旨，这些使它历经百年而不衰，并始终保持在丝绸业及手工缝制行业的领先地位。

中国丝绸第一品牌
中国商业名牌管理委员会

中国丝绸第一品牌

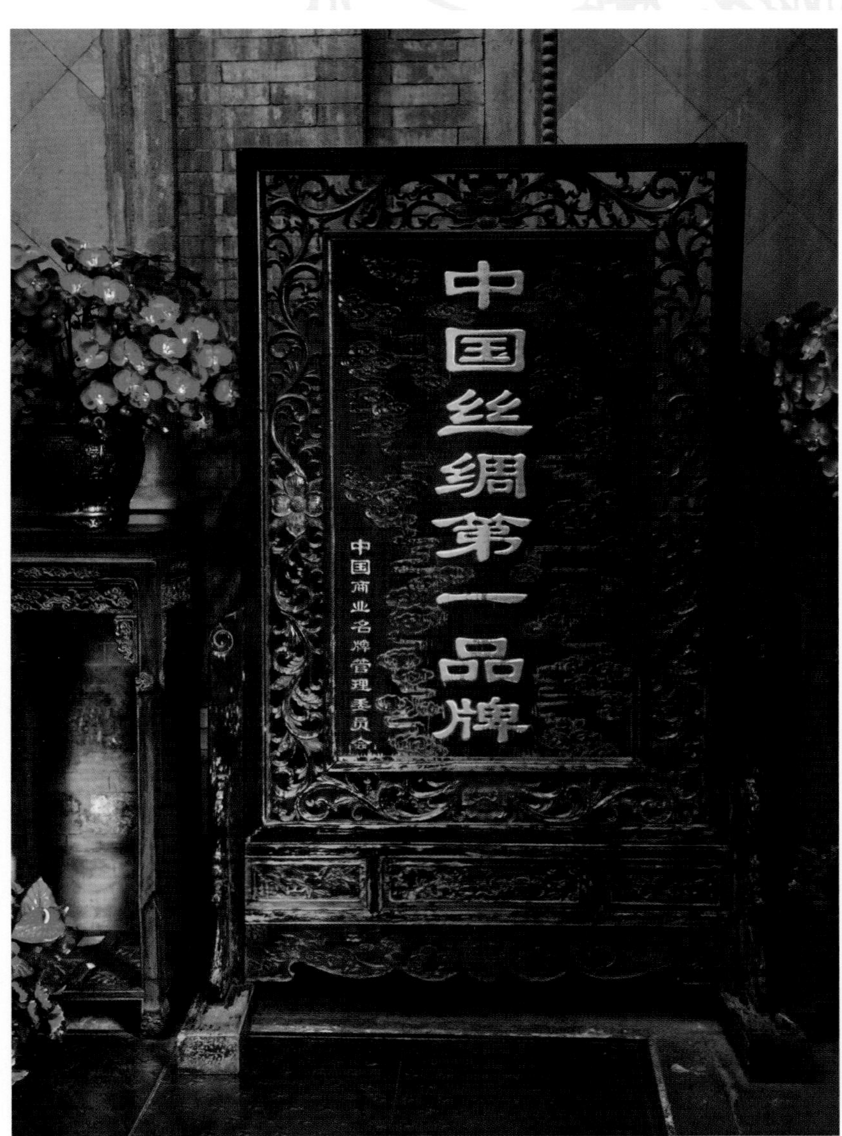

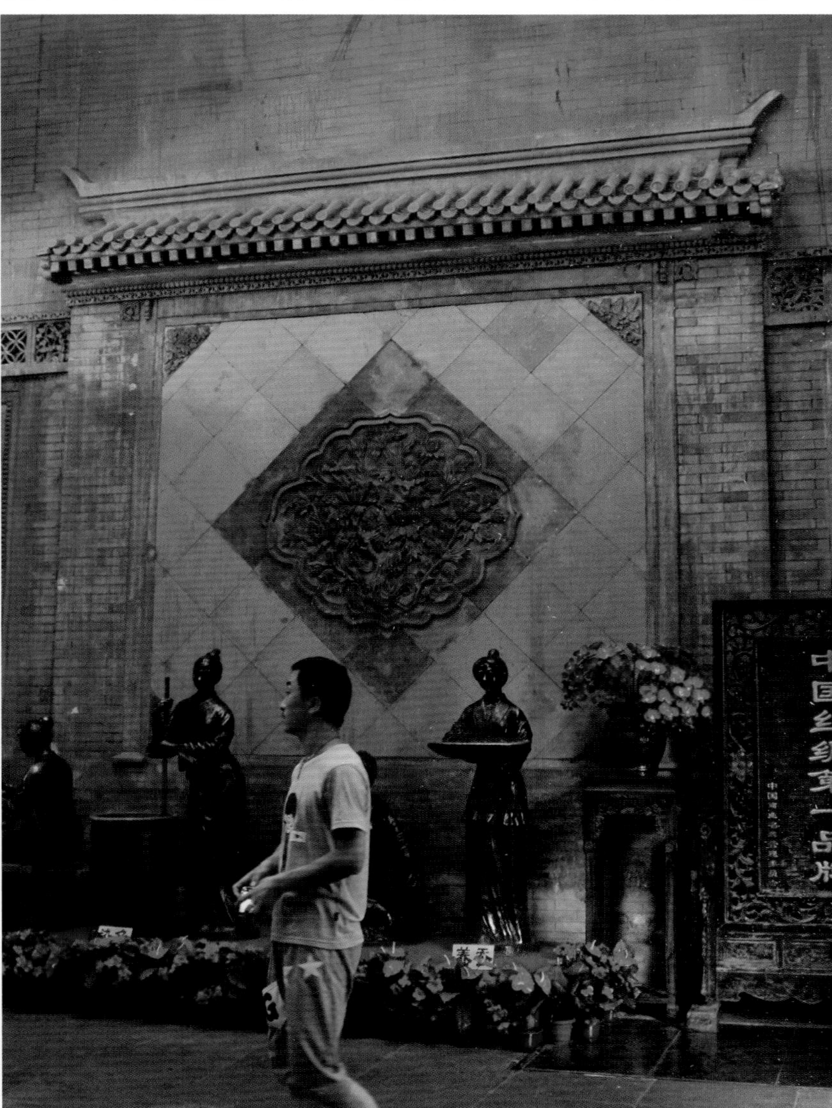

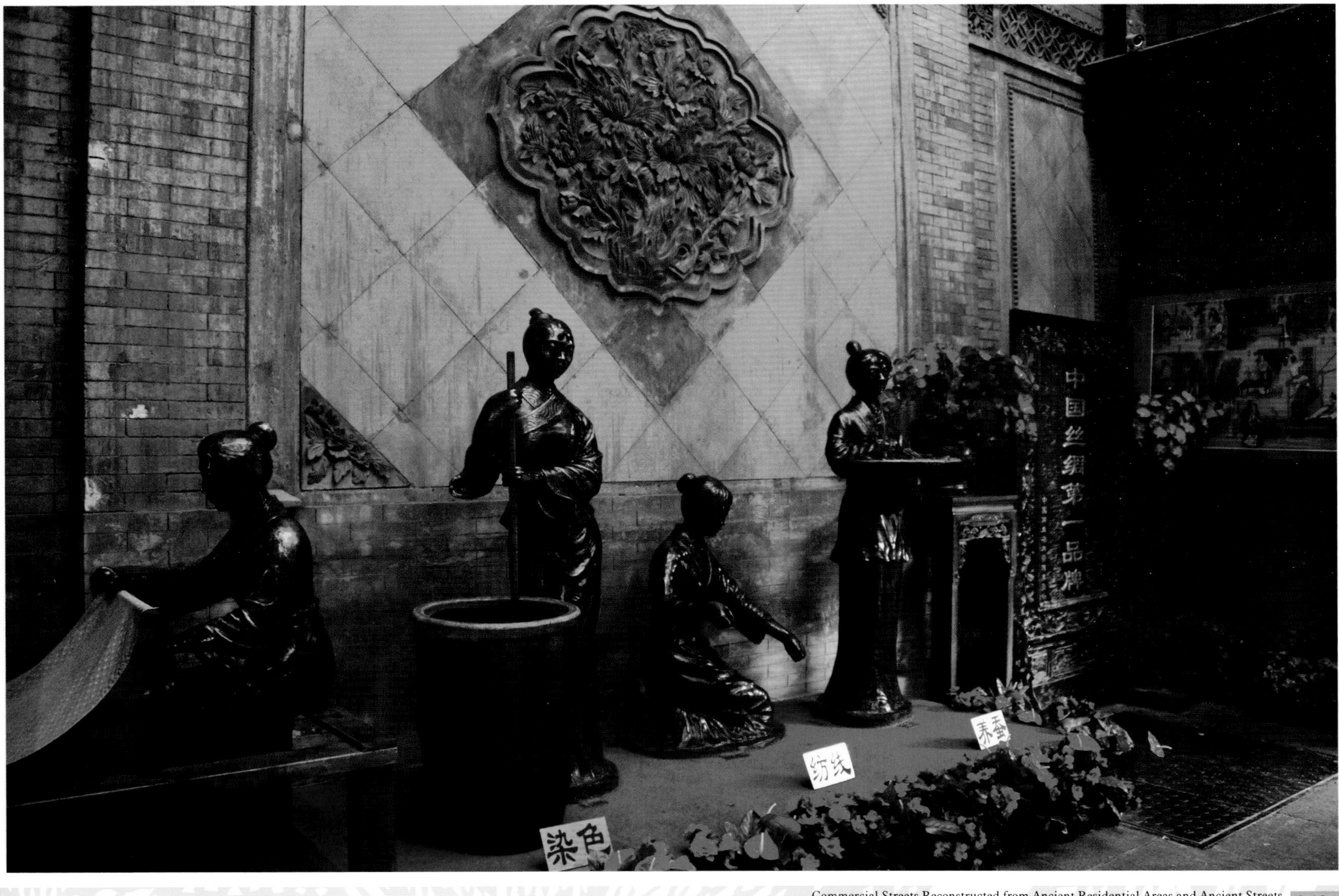
中国丝绸第一品牌
养蚕
纺线
染色

现场手工制作蚕丝被
现场手工制作蚕丝被
玉人物
天然桑蚕丝
现场制作
蚕丝被
真价实

Xiangyi hao Silk

Xiangyi hao Silk was founded in 1896 by Feng Baoyi, a descendent of famous silk merchant Feng family in Hangzhou, and Zhang Xiangzhai, a eunuch manager under the leadership of Empress Dowager Cixi.

Because of its founders, Xiangyi hao Silk specialized in producing court clothes for Empress Dowager Cixi and ministers at the beginning. Later, it also sold tribute silks and satins to the common people and became popular in the folk. During the late Qing Dynasty and early Republic of China, Xiangyi hao Silk became one of "Eight Xiangs".

祥义号

祥义号创立于清光绪二十三年（1896年），由当时杭州著名丝绸商贾世家冯氏家族传人冯宝义与慈禧手下的太监总管“小德张”（本名张祥斋）联合创办。“祥义”二字分别取自两人的名字，寓意“天降祥瑞”“恪守信义”。

祥义号因为创办人的缘故，制衣业务遍布清朝宫廷内部，当时慈禧的衣服及王公大臣们的朝服皆由它制作，做工精美，质量上乘，深得京城达官显贵的喜爱。后来“小德张”说动慈禧，同意宫内绸缎贡品折合银两当作加工宫服的费用，从此祥义号对外经营宫内绸缎贡品，把皇宫丝绸贡品引入民间，受到民众欢迎，清末民初，祥义号成为“八大祥”之一。

Traditional Chinese Medicine 中药类

Tongrentang

Tongrentang which was founded in 1669 by Yue Xianyang is an old and famous traditional Chinese medicine drug store for hundreds of years. The products of Tongrentang have been well-known throughout the country, for its exclusive prescriptions, superior materials, exquisite workmanship and effective treatment.

同仁堂

同仁堂创立于1669年，创始人是乐显扬，至今已有300余年历史，是中国国内久负盛名的中药老字号。同仁堂供奉祖先的牌位前有一副对联大抵能说出同仁堂历时数百年而不衰的秘诀，“修合无人晓，存心有天知”这与同仁堂人恪守的“炮制虽繁必不敢省人工，品味虽贵必不敢减物力”的古训相一致，这种自律精神造就了同仁堂制药过程中的小心翼翼、精益求精的严细精神，它的产品以“配方独特，选料上乘，工艺精湛，疗效显著”而闻名四海。

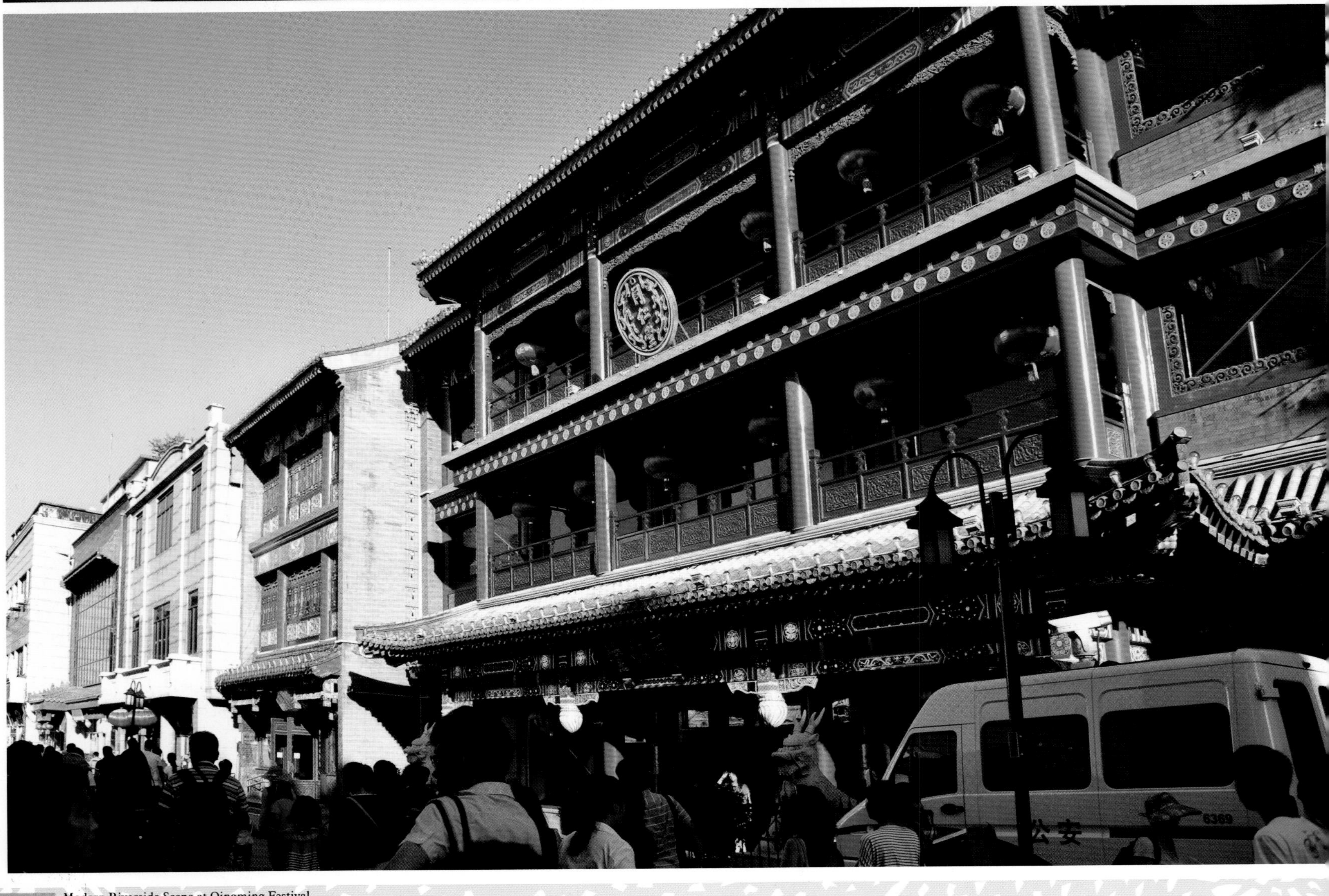

同仁堂文化长廊
中药饮片精品店
同仁堂医館

Jewelry 珠宝类

Yunnan Mineral Jewelry

Yunnan Mineral Jewelry is a branch store of Yunnan Mineral Jewelry Co., Ltd. Yunnan Mineral Jewelry Co., Ltd. which was founded in 1984 is the only jewelry enterprise of Mine Bureau of Yunnan Province, as well as the only jewelry enterprise undertaking exploring, mining, processing and selling throughout the country. Now its main products include jades, gold, platinum, diamonds, gem mounting accessories, silver accessories.

云地矿珠宝

云地矿珠宝是云南地矿珠宝有限公司下的分店，其公司创立于1984年，是云南省地矿局唯一的珠宝企业，也是全国唯一一家探、采、工、贸一条龙的珠宝经营企业。公司现在的主营产品有翡翠、黄金、铂金、钻石、宝石镶嵌饰品、K金、银饰品等。

Leisure 休闲类

Daguanlou Cinema

Daguanlou Cinema was once named Ma Siyuan Teahouse, and was mainly selling tea and offering drama for free. In 1905, it was renamed Daguanlou, and was one of the places that showed films in south Beijing.

After China began to make film, the operator of Daguanlou Cinema bought photographic equipment to begin filming. The first theatrical film *Mount Dingjun* filmed by the operator was first released in Daguanlou Cinema and caused a sensation in Beijing, therefore, Daguanlou Cinema became famous since then.

大观楼影戏院

大观楼影戏院在早期名叫马思远茶楼，以卖茶为主，看戏不要钱，1905年改称大观楼，是当时南城最早的电影放映场所之一。1902年改为大观楼影戏院，根据旧时的报纸，该戏院放映的第一部电影是外国的《麻疯女》，当时引起了很大的轰动，当时播放的电影主要以国外的滑稽片和外洋风光片为主。

到国内开始拍电影后，当时戏院的经营者任景丰添置了摄影器材，开始摄影业务。他所拍摄的我国第一部戏曲片《定军山》在大观楼影戏院首映，轰动了京城，大观楼影戏院从此声名鹊起。

中国电影诞生地
大观楼影城
老电影茶座
方寸斋

China Lane, Chengdu
成都宽窄巷子

街区背景与定位
Street Background & Market Positioning

History 历史承袭

In 1718, the government of Qing Dynasty sent a troop of 3,000 officials and soldiers to Shao City to suppress the rebellion in Tibet. After the war, more than 1,000 soldiers were left here and they were barred from involving in any business activities, but just relied on the jousting bonus to live on. Now, there are only two alleys remained in the city.

In June 2008, China Lane came to the end of its three-year reconstruction work. The new China Lane comprises of 45 building blocks, including courtyards, garden houses, courtyard-style boutique hotels, etc.

清康熙五十七年（1718年），准噶尔部窜扰西藏，朝廷派三千官兵平息叛乱后，选留千余兵丁永留成都并修筑满城——少城。清制规定森严，满蒙官兵一律不得擅离少城参与商务买卖，只能靠每年少城公园（今天的人民公园）春秋两季的比武大会，论成绩优异领取皇粮过日子。风雨飘零，如今的少城只剩下宽窄两条巷子。

2008年6月，为期3年的宽窄巷子改造工程全面竣工。修葺一新的宽窄巷子由45个清末和民国初期的四合院落、兼具艺术与文化底蕴的花园洋楼、新建的宅院式精品酒店等各具特色的建筑群落组成。

Location 区位特征

China Lane is located on the west of the city center, about 1,000 m from the west of the Tianfu Square. It is under Qingyang District's jurisdiction, and stands right in one of Chengdu's Historic Conservation Areas.

成都宽窄巷子位于成都市中心区以西，天府广场西侧约1 000米，属青羊区所辖，在成都三大历史文化保护区之一——宽窄巷子历史文化片区内。保护区北以泡桐树街为界，南以金河路为界，东以长顺上街为界。

Market Positioning 市场定位

The market positioning of China Lane is to form a mixed-use cultural commercial street of distinct regional features and Sichuan cultural atmosphere and develop a new city parlor with cultural deposits of old Chengdu city well reserved, on the basis that the original buildings in the city are well protected.

项目在保护老成都原真建筑的基础上，形成以旅游、休闲为主，具有鲜明地域特色和浓郁巴蜀文化氛围的复合型文化商业街，并最终打造成具有“老成都底片，新都市客厅”内涵的“天府少城”。

Development Concept 开发理念

The program seeks to combine historic conservation area with modern business to find a successful development mode which takes "Chengdu Lifestyle" as its main line. It aims to gather folk life experiences, commonweal exhibitions, high-quality restaurants, courtyard hotels and, so as to develop a courtyard scene commercial street as well as a recreation center of old Chengdu memory.

寻求历史文化保护街区与现代商业成功结合的经营模式，以“成都生活精神”为线索，形成汇聚民俗生活体验、公益博览、高档餐饮、宅院酒店、娱乐休闲、特色策展等业态的“院落式情景消费街区”和“成都城市怀旧旅游的人文游憩中心”。

Street Planning 街区规划

General Layout

There are three commercial streets interconnected to help realize interaction between consumers and commercial activities. Elsewhere, there are two squares on both ends of the Kuanxiangzi Alley to gather popularity and provide promotional venues. Each commercial street is about 400 meters. Those in Kuanxiangzi Alley span 6-7 meters; the others in Zhaixiangzi Alley are 4 meter in width.

Commercial Planning

To highlight the market positioning of fashion life block and experience block overall, the program has the Kuanxiangzi Alley positioned as "Leisure Life", Zhaixiangzi Alley as "Slow Life", and the Jingxiangzi Alley as "New Life", so as to attract all kinds of businesses and organize them with a clear hierarchy. The protection of cultural deposits of old Chengdu and integration of Western culture help realize a combination of the East and West and endow the old lanes with new connotation.

All buildings center on the lanes in courtyard form. Most buildings are two-storey in height, while some courtyard-style hotels are three-storey buildings. All spaces are comfortable in levels, and there is a harmony relationship among the building height, lane width and courtyard scale. From the streets, lanes to household vestibules, the spatial level is clear.

Restaurants and tea houses are spread in the Kuanxiangzi Alley, while Western restaurants and cafes are widely found across the Zhaixiangzi Alley. Bars and night clubs are gathered in Jingxiangzi Alley. There is a clear time sequence among them.

总体布局

项目由3条步行街组成，中间用通道连通，实现人气、商气的互动。另外，在宽巷子两端分别设置了东广场和西广场，起到聚集人气、提供促销场地等作用。步行街长约400米，宽巷子宽度以6～7米居多，窄巷子宽度为4米，街巷距离适宜集聚人气。

商业规划

整体上，为突出项目“时尚生活街区”和“体验街区”的定位，将宽巷子定位为“闲生活”，窄巷子定位为“慢生活”，井巷子定位为“新生活”，既便于吸引各种业态的进驻，又使得整个项目层次分明；既保留了老成都的文化内涵，又融入了西方文化，在实现中西合璧的同时给老巷子赋予了新内涵。

在建筑产品上，整个街区以巷子为中心，以院落为单位，局部2层，宅院式酒店以2层为主、局部3层。空间层次舒适，建筑高度、街巷宽度、庭院宽度和谐，由街到巷再到门厅，院落的空间层次清晰。

在业态搭配上，宽窄巷子以餐饮、茶馆为主，窄巷子以西餐、咖啡为主，井巷子则以酒吧、夜店为主。业态分布体现了时间序列的层次性，即人们上午可以在宽巷子体验原真成都生活，下午可以到窄巷子品味午后时光，华灯初上后可到井巷子体验成都的阑珊夜色。

Street Design Features 街区设计特色

Architecture

All buildings in the program have retained folk house style of Qing Dynasty, and herring-bone, a common form in north China road system, is main feature of the streets. Traditional buildings along the streets are well preserved, and the gate of courtyards is the most common scenery.

Landscape

The landscape design of the streets, lanes and squares reflects the traditional deposits of the city of thousand years of history. In the courtyard design, attention is paid to residence culture of "life fun" and "multiple functions", so as to provide users with elegant, understated but luxury life experience. The plain and tranquility of streets and lanes and sophistication and luxury of courtyard form shark visual and functional contrast. They reflect different historical scenes and realize both the protection and development of the site.

The Kuanxiangzi Alley is graceful, with roads regularly laid with black stone plate, and black bricks of the same color running throughout each courtyard. Horizontal stone bars are used to enrich the pavement of the roads lined by hitching posts, archaic street lamps, mighty stone lions in front of each courtyard, ancient wells under old trees, etc. Planting beds are paved with natural stones, together with dense woods, wild shrubs and thriving vines, to form a green alley.

The Zhaixiangzi Alley is free and cozy. The black stone roads feature regularly lain stones and mosaics. The artistic facades and smooth stone balls under the eave form strong visual contrast. Planting beds are free and abstract in forms and mingle with the road pavements. On two sides of the road, stone benches and stools provide very convenient places to rest.

design inspiration derived from the allusion of the soldiers. The square includes a performance stage, auditorium areas, bronze statues, arrow targets and an archway. The performance stage can be used for Sichuan drama and storytelling performance venue, and the archway, as a symbolic building, will serve as the linker between the Kuanxiangzi Alley and the Zhaixiangzi Alley.

The landscape designed at the entrance of Jinhe Road also combines the two alleys together. On each end of the road, a cobblestone rockery reproduces the terrain of Western Sichuan region and enhances the continuation of the alleys on the road.

Lighting

To create tranquil, leisure and slow atmosphere, light intensity in the program is under strict control. Warm colors of low color temperature are used in courtyards to shorten the distance between people and enhance the hazy feeling. Chinese style facades are equipped with spotlights to highlight door heads, while the rest of the building is hidden in dark. To reflect life breath of the buildings, some lights are embedded within the gaps between the tiles.

Bird-shaped lanterns are used at main entrance, and their design inspiration comes from the bird-breeding habit of Manchu and Mongolian tribes as well as typical Chinese cultural symbols. They are creative and natural that they become an integral part of the environment.

Corrugated lamps are vastly used as wall lamps in the program. Typical in shapes and rich of Chinese cultural elements, they become an eye-appealing landscape on the spot. Through the well designed corrugated structures above the lights, one can see the rooftop structures clearly, and thus the visitors and scenic spot become more intimate.

建筑

项目延续了清代川西民居风格，街道在形制上属于北方胡同街巷，其主要特色为“鱼脊骨”形的道路格局。沿街传统特色立面保存完好，其中以院门的形式最为丰富。建筑作为空间的表皮，是空间历史感的外部表象，通过实体界面的强化，让历史街区重塑出空间的时间厚度。

景观

街巷、广场节点等公共空间的景观设计体现了少城千年的城市传统沉淀。在庭院设计中注重“生活情趣”和“多元繁复”的公馆文化，为内部参与者提供优雅、低调、奢华的生活感受。街巷的古朴静谧和庭院的繁复奢华构成视觉及功能的对比，体现了不同的历史片段，实现了保护和开发的兼顾。

宽巷子雍容大气，路面以整齐的青石板铺装为主，采用同一色调的青砖串联宅院内部交通。以横向的条石丰富路面铺装形式，道路两侧重复出现的拴马桩、形式古朴的路灯、宅院门口威武的石狮子、苍老的大树下古色古香的水井等等，把游人带回到古老的少城。种植池多由自然的块石砌筑而成，茂密的绿荫、自由式生长的灌木、随处可见的爬藤组成了一条绿色的街巷。

窄巷子体现自由惬意。青石板路面由规整和碎拼两种方式组合而成，更多风格的景观元素掺杂其中，充满装饰艺术风格的建筑立面造型与屋檐下光滑的石球形成强烈的对比，老巷子的魅力在这种对比中展现得淋漓尽致。种植池设计成自由、写意的形式，与道路铺装间没有明显的分界线，道路两侧上的石制座凳为游人提供了便利的休憩设施。

东入口广场设计为“演武广场”，创意构思来源于少城官兵的典故。演武广场的设计元素由演武台、观演台、情景铜人雕塑、射箭靶、城守府门楼构成，其中演武台可作为川戏、评书的表演台，城守府门楼是一座象征性的建筑，作为宽巷子的入口与宽巷子紧密结合。

临金河路入口景观设计把宽巷子和窄巷子的入口结合起来设计，两个入口各放一座川西风格的假山，中间用川派特色的鹅卵石堆砌而成，模拟川西山脉山势的微地形小山包，增强宽窄巷子在临金河路的延续性。

灯光

由于街区的氛围特点是静、闲、慢，因此，整个照明体系严谨并严格控制亮度。街区中的院落在色温上为低色温的暖调，不仅能有效拉近人与人的距离，而且还能增加环境朦胧感。房屋则采用不勾勒建筑轮廓、不照亮瓦楞的手法，中式建筑立面使用投光灯，在有特色的院落门头给予突出表现，使之成为故事节点，增强街区的叙事性。同时，为体现房屋的生活气息，在有瓦花的位置做内透照明。

鸟笼灯是街区主要入口的灯型，设计围绕满蒙旗人的养鸟习惯和中国文化的典型符号展开。灯形有新意、不夸张、不夺目，与宽窄巷子的环境水乳交融，成为环境中不可或缺的部分。

瓦楞灯是街区中多处使用的壁灯，造型经典且富含中国文化元素，成为了景区中吸引眼球的景观之一。同时，对灯具上方的“瓦楞”部分做精心处理，把街区中不能看清楚的房屋顶部结构展现在游人面前，拉近游人与景区的距离。

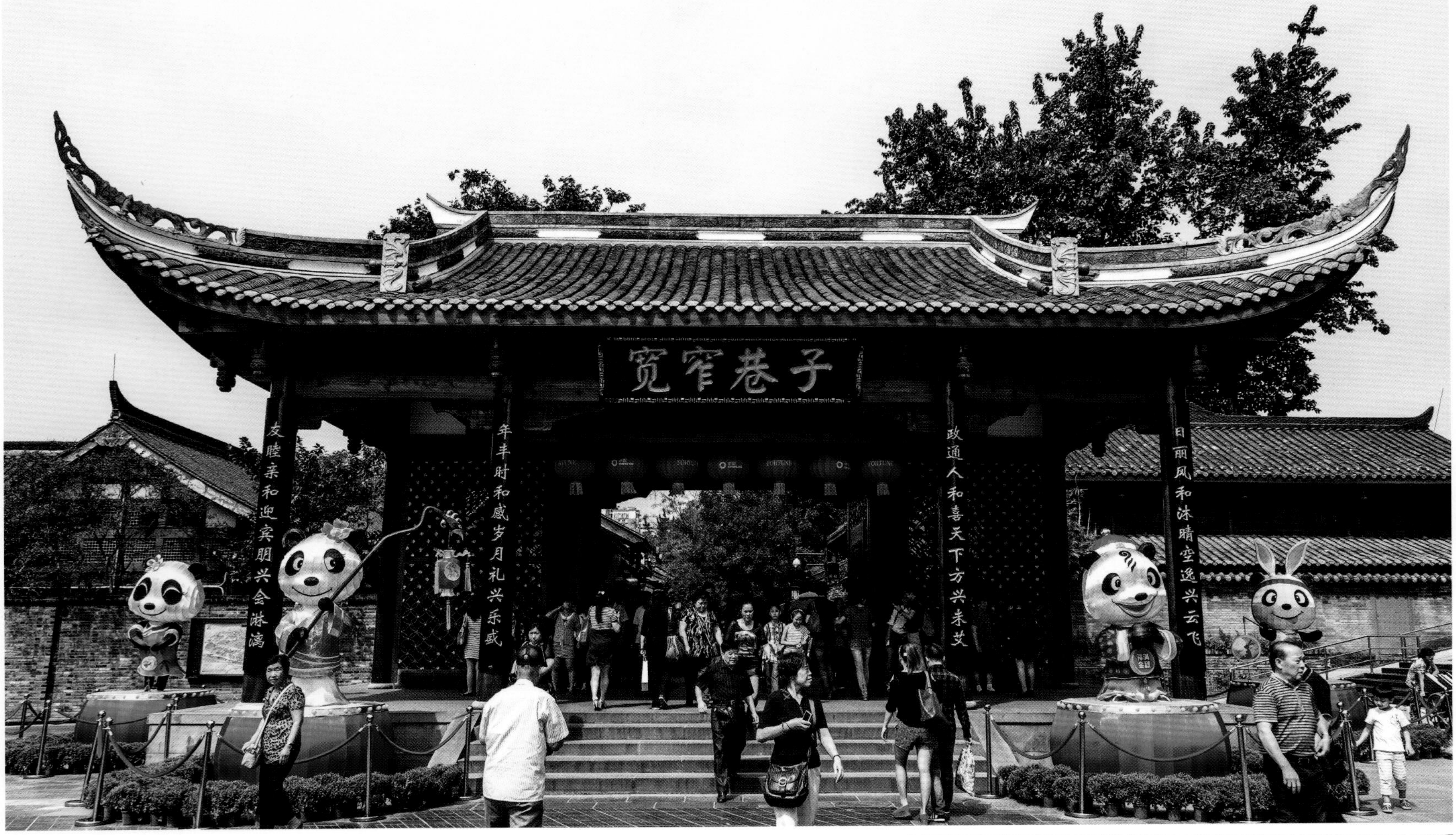
宽窄巷子
友睦亲和迎宾朋兴会淋漓
年丰时和感岁月礼兴乐盛
政通人和喜天下方兴未艾
日丽风和沐晴空逸兴云飞

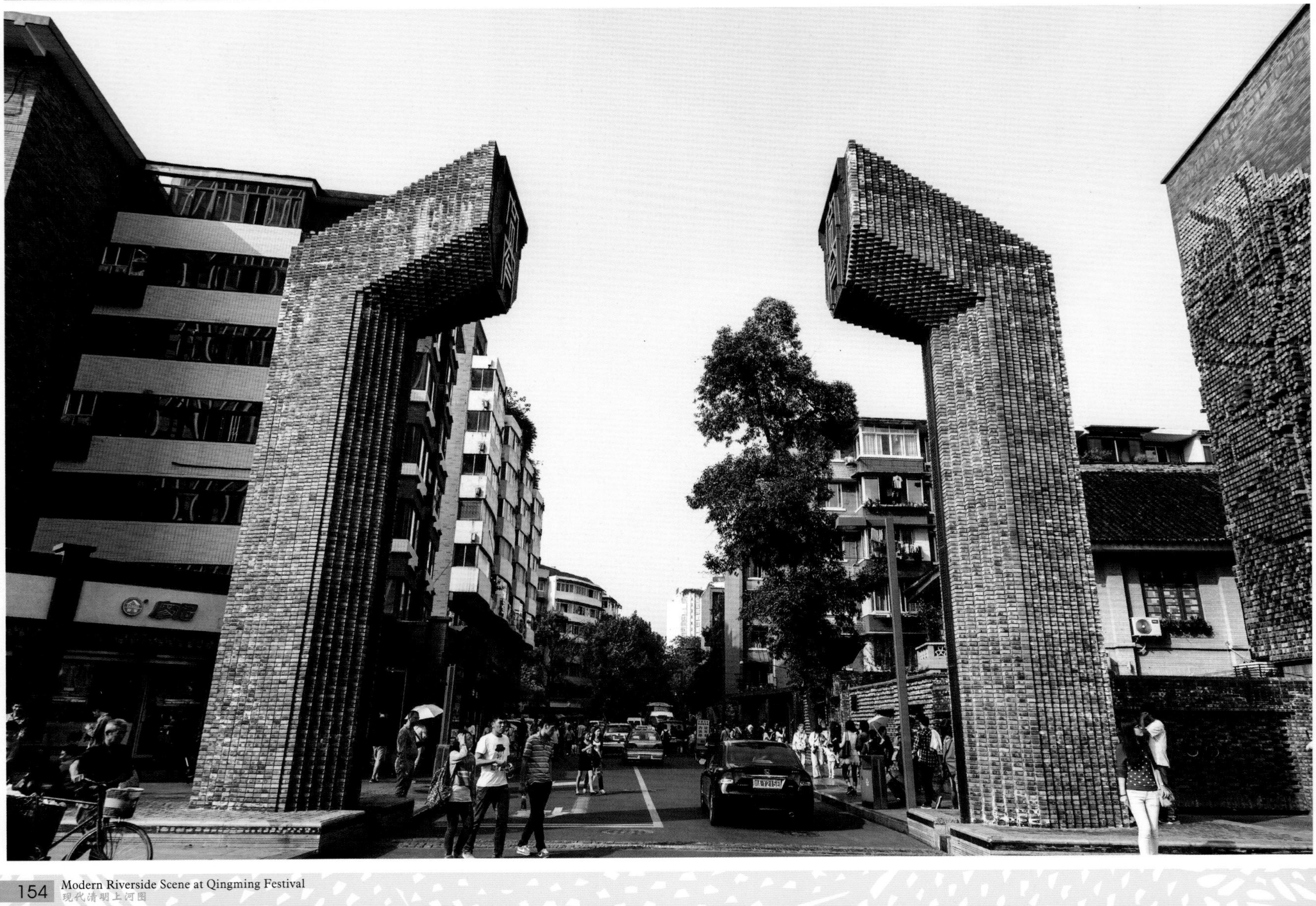

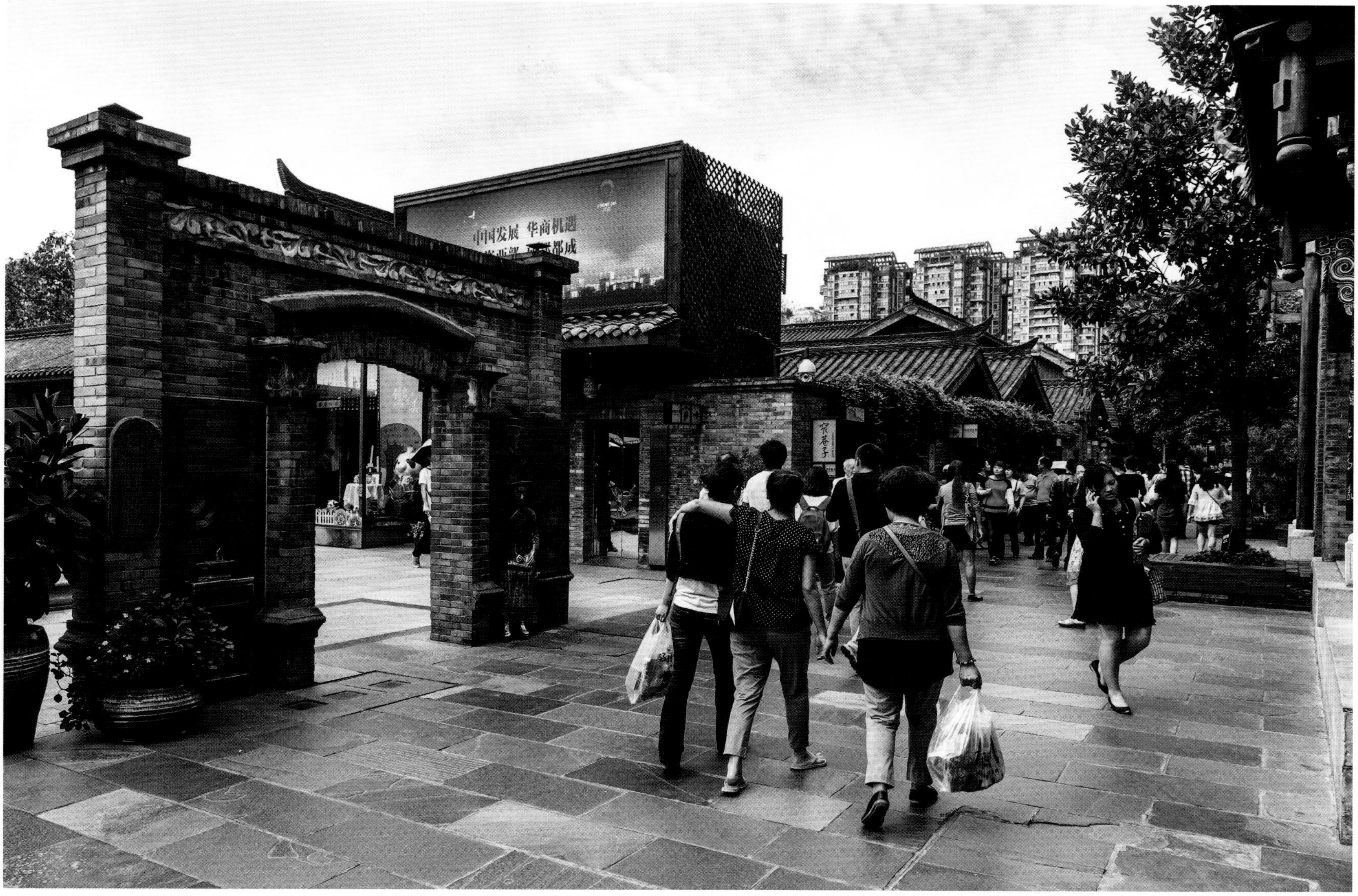

寄给
未来

漆重門

商业业态与运营

Commercial Activities & Operation

Major Commercial Activities 主要商业业态

There are many leisure activities in the Kuanxiantgzi Alley, including boutique hotel, private dining, fork catering, leisure tea house, inn, enterprise club and SPA. The Zhaixiangzi Alley is composed of Western restaurant, fast food shop, cafe, theme cultural shop, and so on, for exquisite life experience. The Jingxiangzi Alley is a stylish and lively entertainment area covering bar, night club, dessert shop, tiny specialty store, and fast food shop.

在业态上，宽巷子形成以精品酒店、私房餐饮、民俗餐饮、休闲茶馆、客栈、企业会所、SPA为主题的情景消费游憩区。窄巷子形成以西式餐饮、轻便餐饮、咖啡、艺术休闲、健康生活馆、特色文化主题店为主题的精致生活品味区。井巷子形成以酒吧、夜店、甜品店、婚场、小型特色零售、轻便餐饮、创意时尚为主题的时尚动感娱乐区。

Operation Measure 运营措施

Import and Export

As for export, attention is paid to overseas marketing and marketing tools innovations. Through cooperation with well-known international brands, and establishment of overseas marketing centers, the program aims to improve the international awareness of the city and promote it to the world, making it a pioneer in urban marketing. For import, Pandahome, a web portal for tourism promotion, was established in 2008. In 2010, Chengdu and China Travel Service in Hong Kong established a cooperative relationship. Later, the program found more cooperative partners such as Disney, Google and Pepsi to promote local cultural tourism development and commercial brands.

Scientific Program Management Method

To ensure professional management and operation of the program, local government established several related departments and rules to strengthen the comprehensive law enforcement and integrated management.

Moreover, Chengdu Culture & Tourism Development Group also specially set up Chengdu Culture & Tourism Wealth Management Limited Company to be fully responsible for the project investment, operation and management, to center on market to run and promote the program on the city marketing platform and height, adhering to a high-standard, profession management philosophy to introduce shopping mall management mode, relying on overall marketing to sign strict business management contracts with merchants, merge business scope, decor, stalling specifications, service etiquette for uniformed management, and establishing the system of rewards and punishments to constrain and encourage all business. At the same time, the group also carried out many targeted advertisings, promotions and events and successfully created many famous brands such as the Forever Young Concert, Kuanzhai Tea Party, Kuanzhai Music Season, etc, to improve the awareness and reputation of the program.

“走出去”与“引进来”

以境外营销为重点，创新营销手段，通过与国际知名品牌企业合作、在境外设立营销中心等举措，全面提升成都的国际知名度，将成都推向世界，成为城市营销的先锋力量。2008年创立了成都旅游营销的网络门户“Pandahome”；2010年与香港中国旅行社合作成立成都旅游香港运营中心，向全球营销成都文化旅游产品，提供旅游信息服务；先后与迪斯尼、谷歌、百事可乐等国际巨头合作，推广成都文化旅游和成都文旅品牌。

科学的项目管理方法

为确保宽窄巷子的专业化管理和运营，青羊区政府成立了分管副区长任组长、各职能部门参与的“宽窄巷子历史文化保护区管理委员会”，由街道与文旅集团资产管理公司负责人组成管委会办公室负责日常管理，并制定了《宽窄巷子历史文化保护区管理办法》，强化综合执法和综合管理。

成都文旅集团专门成立了文旅资产运营管理公司，全面负责宽窄巷子项目招商、运营和管理，以市场为中心，站在城市营销的平台和高度上进行运营和推广。秉承高水准专业运营的管理理念，引进商场式管理模式，依靠市场整体营销，与商家签订严格的商业管理合同，对经营的业态范围、装修风格、外摆规范、服务礼仪等进行统一管理，并建立奖惩制度，进行约束和激励。同时，有针对性地开展街区的宣传、营销、活动策划等工作，根据商业业态开展主题活动，成功打造了井巷子创意市集、“永远年轻”跨年摇滚音乐会、宽窄茶会、宽窄音乐季等自有品牌，提升了宽窄巷子的知名度和美誉度。

Food & Beverage 餐饮类

Three Bricks Restaurant

Three Bricks Restaurant of courtyard buildings, sunshine, lifting wine cellar, plain black roof tiles, grand beams and eaves restores the history and vigor of culture in the bustling Chengdu city. Its three plain colors will bring you back to the simple life. It is an ideal place for family reunion, dating, business talk, friend gathering, working lunch and community meal.

三块砖餐厅

合院建筑、中庭阳光、升降酒窖、古朴青瓦、雄瞻梁檐……在繁华喧嚣的都市，“三块砖”还原了历史和人文的活性，在三种质朴之色中寻找返璞归真的生活况味。特色菜式有夫妻肺片、私味酱鸭、春蚕吐丝、蘸酱黄瓜、麻辣拌鸡、养生鲜菌汤、茉莉花松露蒸蛋等，是家庭聚会、情侣约会、商务洽谈、朋友聚会、工作午餐、大型聚餐的好去处。

Jinshan Seafood Hot Pot Restaurant

The restaurant for fish hot pot of Sichuan taste renovates the original courtyards and subdivides them into many small private dining rooms with elegant and traditional environment. The river fish hot pot is tasty.

尽膳河鲜馆

尽膳河鲜馆是一家以川味鱼火锅为主的餐厅，结合原来的巷院修旧如旧，分隔为一间间小包间，环境幽雅，很有传统特色。火锅以当地河鱼为卖点，肉质细腻，骨刺较少，放入沸腾的红汤中一涮，令人拍案叫绝！

Elegant Courtyard Cafe

Elegant Courtyard Cafe is a club renovated from a ancient private courtyard. With new architectural and interior designs as well as garden landscape, it provides a excellent communication platform for exclusive classes to exchange culture, studying literature works and launching products. It is also a comfortable, tranquil, elegant space for leisure and social communication.

里外院咖啡厅

里外院将古老的私人院落改为里外院建筑会所，以全新的建筑设计、空间设计、园林景致作为视觉呈现，建造出一个文化交流、著作讲习、产品发布等专属圈层的交流平台。此外，里外院也以普洱茶茗、原生态美食、小众派对、高规格会议接待为媒，以极具品位的艺术氛围，创造了一个舒适、幽静、雅致的休闲和社交空间。

Chengdu Image Snack Restaurant

This restaurant comprises of five storeys. The ground, second and third floors house dining halls and 15 private dining rooms. There are archaic wooden furniture, carved doors and windows and old Chengdu scenes painted on the walls. The first floor underground serves as tea room and drama performance stage of ancient Chengdu beauty. The entire building is centered on a patio which can protect the interior from the storm and bring in ample natural daylighting. Therefore, the Sichuan drama performing team in the restaurant called themselves "narrow patio drama troupe". The unique operation concept will bring all dinners with sweet homey feelings.

成都映象小吃店

本店建筑共有5层，平层及二、三楼设用餐大厅及15个大小不同的包间，内有古色古香的木质家具、雕花门窗，墙上还有老成都旧时场景绘画作品，负一楼是还原老成都风貌的茶馆和戏台。整栋建筑的正中有一个上下贯通的转角式天井，可挡风雨，而阳光却能从天井中射下，也正因这个原因，成都印象的川戏表演团名为"窄天井剧团"。成都印象以"重温往日成都巷陌，回归道地川菜本味"为经营理念，让食客重温内心深处故土家乡的温暖亲切。

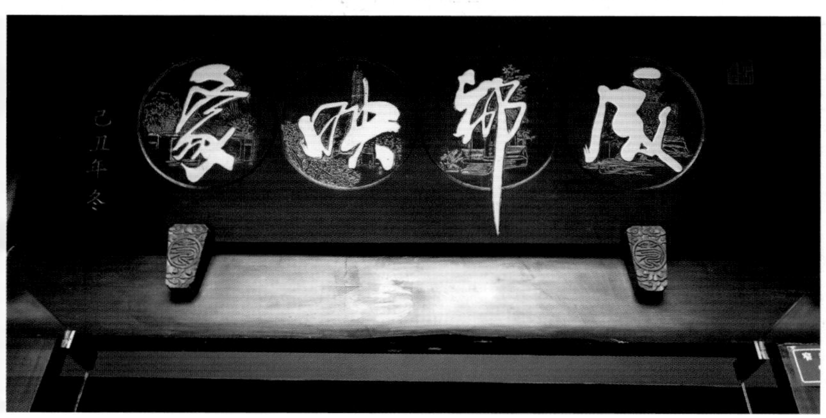

Shangxi Restaurant

This bid courtyard of old black brick walls, moon gates, tree houses only has a dozen of dinning rooms. It is said that the predecessor of the courtyard is an elementary school in Qing Dynasty.

上席餐厅

青砖老墙、老门洞、老树，偌大的院子里只有十几间包房。院落据说是曾经的清朝的一座小学，坐在院子里喝茶会友，很有些古意。包间名取自《周易·雅致》，菜品讲究，由烹饪大师亲自调理。

Longxuan Private Kitchen

The name of the restaurant comes from the old house of Nian Gengyao, a famous general in Qing Dynasty. The architectural appearance has the courtyard style of Qing Dynasty completely preserved, while the interior design combines traditional Chinese decor with modern Western forms and arts perfectly. Main cuisines include Tanjia cuisine and new Sichuan cuisine.

龙喧私房菜

龙喧，原为清代名将年羹尧于康熙五十七年间驻扎少城时的旧居。建筑外观完全保留清代宅院建筑风格，室内设计把中式传统装饰和西方现代造型艺术完美结合，独具匠心的装修风格更显龙喧雅致高贵的宫廷气派。主营菜系为官府菜和新式川菜。菜品均选用上等食材，由国内顶尖厨师精心烹饪，南北膳食风味的碰撞和融合让食者体验绝佳的味觉享受。

Desir French Restaurant & Lounge

Desir French Restaurant & Lounge will offer you authentic French cuisine, desserts, wine and coffee.

古魅法式餐厅

古魅法式餐厅醉心于追求艺术文化的内涵和精致的生活品位，主营正宗的法国大餐、甜点、红酒和咖啡。此外，还有动听的法国香颂做陪衬，使食客在品尝美食的同时真正了解法国各地区不同的风土人情。

Three Ears Hot Pot Restaurant

Three Ears Hot Pot Restaurant is a chain hot pot restaurant. Currently, the brand has 46 directly-managed branches and chain stores around the country.

三只耳

三只耳火锅是一家连锁火锅店，目前在全国已经有46家直营店和加盟连锁店。三只耳首创的冷锅鱼，其汤料口感独特，麻、辣、鲜、香、嫩、滑、爽，层次分明，在原料选择、工艺创作上自成一派。

Leisure 休闲类

Wang Hong Ji Custom Shop

This space invested and designed by Wanghong Culture Investment Limited Company houses cultural design, philosophical design, innovative design, as well as national historical relics.

王红彐定制

王红彐定制是王红文化投资公司投资并设计的五行主题空间，内有文化设计、哲学设计、创意设计，并且展示国家级文物。

Zhuyeqing Tea Shop

Zhuyeqing Tea Shop is specialized in manufacturing and selling Zhuyeqing tea series as well as many other tea brands such as Lun Dao, Piao Xue and Baoding Xueya.

竹叶青茶叶店

竹叶青茶叶店以生产、销售"竹叶青"系列名茶为主，拥有论道、竹叶青、飘雪、宝顶雪芽等茶叶品牌。其中"竹叶青"是全国十大名茶之一、"四川名牌"产品和四川省著名商标，"竹叶青"为陈毅元帅1964年在峨眉山万年寺与得道高僧对弈品茗而得名。

Ke Ju Tea Shop

Ke Ju Tea Shop is originally the name of a tea lover's study and means no complaining and regret. Pushing the wooden door of Ming and Qing dynasties open, visitors will find a bar counter with carvings for reception, and various teas, tea wares are also displayed here.

可居

可居，与茶有关，源自一位爱茶者的斋名，意为“无须抱怨、亦不需懊悔”。推开明清木门，即见一个镶着雕花、映着灯光的吧台，是存茶、取茶、接待之处，亦把茶、茶具置于此展示。

Qun Ying Hui Gift Store

This store is mainly specialized in exquisite gifts of various styles and prices, such as folk style earings, bracelets and other accessories.

群英绘

群英绘主营不同风格种类与价位的精美礼品，如民族风格耳环、手镯、饰物等。

A panda Creative Living Gallery

Apanda Creative Living Gallery is a specialty store committed to sharing the most fresh creative life information, providing the most popular creative life experience, and transmitting the most relaxed and happiest life attitude.

熊猫创意生活馆

熊猫创意生活馆是一家致力于“分享最为新鲜的创意生活资讯、品味最为流行的创意生活、传递轻松快乐的生活态度”的特色店铺。

Crafts 工艺品类

Lianhuazi Featured Shop

Lianhuazi Featured Shop mainly trades in handcrafted bronzes, burners, incenses, Fengshui accessories, and so on. There are also bracelets and Buddha beads made of various natural stones.

莲花子

莲花子是一家特色小店，主要经营手工铜器、香炉、香薰、风水摆件等工艺品。这些工艺品做工精细，据说中国目前最好的仿古香炉都能在这家不起眼的小店里找到。此外，这里还有各种天然宝石做成的手珠、佛珠。老板是信佛之人，闲时到处淘宝，为了几尊佛像、几串老珠子，他最远到过印度、尼泊尔，跋山涉水亲自将其淘回来，然后结合他独到的设计灵感，把传承千年、辗转千里的珠子搭配成一串一串充满灵性和美感的手珠、佛珠，等待那个有缘的人。

Travel Culture Creative Shop

The irregular connection of stone pavement and garden vegetation in Travel Culture Creative Shop creates a well-proportioned artistic effect of rich connotations. The shop reflects the profound cultural deposits of Chengdu city as well as its gorgeous history, and more importantly, shows the bright future of the city, through creative design.

行走旅行主题创意店

店内石材、花园式种植等不同饰面无规则衔接设计，形成了错落有致、内涵丰富的艺术效果。店铺体现了成都深厚的文化积淀和辉煌的历史，更为重要的是，通过富有创意的设计，展现了成都光明的未来。

Recreation 娱乐类

Ou Er Bar

Staring from sense experience, Ou Er Bar designs and creates a wide arrange of entertainment, and implants multiple art forms such as music, theater and exhibition. World-class array sound system creates unrivalled on-site effect which in turn becomes a new card of the bars in the Zhaixiangzi Alley.

偶尔酒吧

从感观体验出发，设计与创造了多元化的娱乐方式，并植入音乐、剧场、展览等多种艺术形式。世界级顶尖阵列音响系统，营造无与伦比的现场效果，成为宽窄巷子酒吧的新名片。

天堂眼
swarka
天工物

China red
中国红

茶马江湖

緣覺

藏蕴堂

寬居

老香港茶餐厅

浮廬

Furong Street, Jinan
济南芙蓉街

街区背景与定位

Street Background & Market Positioning

History 历史承袭

With a long history, the first built year of Furong Street is impossible to be verified. However, from the first built year of Confucious Temple we can know that it had a certain scale since North Song Dynasty. During Ming and Qing dynasties, there were government authorities in this street. There were also a few scholars' houses here. The good geographical condition promoted its economic development and cultural undertakings.

Furong Street has wonderful sceneries. Its spring water culture is different from Jiangnan's water flowing beneath little bridges. Furong Street is bustling and integrates primitive simplicity with modern life style after a long history. It is just like Jinan people who keep traditional honesty and sincerity in this fast-paced modern society.

芙蓉街历史悠久，其始建年代已无法考证，但根据学府文庙的始建年代，芙蓉街应在北宋熙宁年间就小有规模。明清时期，这条街道四周多分布政府权力机关，如巡抚院、都司、布政司、贡院、府学衙门等。在这古老的街道中，也有不少学者落户此处，明朝中期诗人许邦才就在芙蓉街附近建造“瞻泰楼”，清代诗人董芸业曾在芙蓉馆中居住。芙蓉街良好的地理环境促进了其经济文化的发展，众多名商富贾开业经营，如瑞蚨祥布店、“一珊号”眼镜店、文升行百货商店等；文化事业也遍地开花，著名的教育家鞠思敏、王祝晨等曾在这里开办教育图书社，知名画家俞剑华、岳祥书也在此处开馆授徒。

芙蓉街风景优美，与江南的小桥流水不同，这里是一种别样的水文化——泉水文化。董芸曾写过一首诗来描绘芙蓉泉的美丽：“老屋苍苔半亩居，石梁浮动上游鱼。一池新绿芙蓉水，矮几花阴坐著书”（《芙蓉泉寓居》）。芙蓉街繁华热闹，历经沧桑，将古朴与现代融为一体，就像济南人，在快节奏的现代社会中生活，又保持着传统的淳朴与敦厚。

Location 区位特征

Located at the center of Jinan, Furong Street is named after the Furong Spring in the street. It is surrounded by ancient buildings and the main road of ancient city, extends from Xihuaqiangzi Street South Gate in the north to the middle Quancheng Road in the south, and its east is adjacent to Xi Lane and west to Gongyuanqianggen Street. This old commercial street is called "the top snack street in Shandong".

芙蓉街位于济南市中心，因街中芙蓉泉而得名，位于珍珠泉群之中，地理位置优越，北起于西花墙子街南口，南至繁华的泉城路中段，东临县西巷，西靠贡院墙根街，紧邻两大府衙和贡院等历史建筑及古城主干道。全街长432米，宽约6米，是济南市的老商业街，是全国最为著名的小吃街之一，被誉为“齐鲁第一小吃街”。

Market Positioning 市场定位

As one of the most famous old streets in Jinan, Furong Street reserves lots of historical buildings including the natural landscape Furong Spring which is in the list of Seventy-two Springs in Jinan. Adopting the method of combing antique mending and modern technology, the developing of Furong Street aims to continue the street's history, improve residents' lives and make full use of the street's economic value on the base of well-reserved historical sites. The street is built to be a multicultural recreational business district that integrates landscapes, residences, business and culture, and satisfies tourists with entertainment, shopping, food and leisure.

芙蓉街是济南最著名的老街之一，保留了大量的古代历史建筑，有名列济南七十二泉之中的天然泉景芙蓉泉，它既拥有历史的厚重，又有自然的灵性，在开发的时候就要以现有保存较好的天然泉景、古街巷、古民居、文物古迹为基础，以延续历史存在，改善居民生活，发挥经济价值为目的，以恢复该街区的历史古城风物、风貌、风情为着眼点，以重现历史古城的多元文化为主格调，采用修旧仿古与现代科技艺术相结合的手法，将该街区建成集泉景、民居、商业、文化、民风民俗等景观为一体，满足市民和游客观光、娱乐、购物、餐饮、休闲等需求的多元文化的中央游憩区。

规划设计特色
Planning & Design Features

Street Planning 街区规划

The planning protection and control area of Furong Street is about 468,100 square meters. This street integrates lanes, springs, historical sites, ancient building and folk customs to construct unique cultures of history, architecture, spring and celebrity. It is widely accepted as the "soul" of Jinan by worldwide tourism experts and Jinan citizens.

The planning of Furong Street puts emphasis on protection and renovation. In 1999, the government of Jinan invited experts of Institute of Architectural & Urban Studies of Tsinghua University and Jinan Urban Planning & Design Institute to plan how to protect and renovate this area. In 2000, the government invested 500 million RMB to build Furong Street into an "ancient city sightseeing district". From 2006 to 2010, Furong Street was gradually reformed from a "snack street" into a "tourism and business street". After Municipal Political Consultative Conference in 2011, the sub-district office of Jinan re-positioned Furong Street as "Cultural tourism featured street" on the base of collecting public opinions.

芙蓉街历史街区规划保护控制面积约46.81公顷，街区融街巷、泉群、名人古迹、古建筑、民俗风情为一体，构成了独特的历史文化、建筑文化、泉水文化和名士文化，是世界旅游组织专家及济南人普遍认可的济南的“灵魂”。

济南市对芙蓉街区的规划以保护整治为主。早在1999年，济南市就开始酝酿芙蓉街和曲水亭地区的保护与整治，为此济南政府邀请了清华大学建筑与城市研究所的专家与济南市规划设计研究院一起对该地区的保护和整治进行规划。2000年，济南市政府投资5亿元将芙蓉街建成“古城游览区”，将芙蓉街打造成为继“山、泉、湖”之后的济南第四大旅游景点。2006年，泉城街道办事处对芙蓉街进行了首期的改造工作，改造路面，拆除违章建筑，修复老字号店铺等。2009年到2010年，政府提出改变芙蓉街“小吃一条街”的角色，将其打造成“旅游商品一条街”。泉城街道办事处修缮了芙蓉街北段沿街建筑物，清理芙蓉泉，对街口牌坊翻新整容。2011年，在市政协会议后，广收民意，泉城街道办事处将芙蓉街重新定位为“文化旅游特色一条街”。

Street Design Features 街区设计特色

Furong Street is named after Furong Spring which ranked forty two in Seventy-two Springs of Jinan. Different from unconstrained Spouting Spring and restrained Pearl Spring, Furong Spring hiding itself in a residence has a beauty of a pretty girl of humble birth.

There are not a few ancient celebrities' houses and historic buildings at the Furong Street, such as the residence of Dong Yun, a poet of Qing Dynasty, and four temples including temple of local god of the land, Dragon Temple, Temple of Guan Yu and Confucious' Temple. The buildings at Furong Street reflect the development of late Qing Dynasty.

When walking on Furong Street, you can smell various kinds of snacks and hear peddlers peddling their wares. Eating different delicious foods from different places, you will enjoy yourself so much as to forget to leave.

芙蓉街由芙蓉泉得名。芙蓉泉在济南七十二泉中名列第四十二，它与趵突泉的豪放不同，也有别于珍珠泉的婉约，它藏身于民宅之中，极具小家碧玉的特色，给人一种“藏在深闺人未识”的感觉，又有“犹抱琵琶半遮面”的风韵。因为芙蓉泉，芙蓉街有了《老残游记》中“家家泉水，户户垂柳”的描述。

芙蓉街中有不少名人古迹、古建筑，清代诗人董芸居住的“芙蓉馆”因其书声琅琅、流水潺潺、垂柳依依、意境优雅而四海闻名。芙蓉街的四庙（土地庙、龙神庙、关帝庙、文庙）也闻名遐迩。芙蓉街区的建筑反映了清末民初的发展变化，中西合璧的建筑保留至今。

芙蓉街是吃货们的天堂。踏在芙蓉街的路面上，鼻中闻的是各色小吃的香味，耳中是路旁小贩的吆喝，口中吃的是来自各地的风味美食，热闹的市井景象，让人流连忘返。

假发

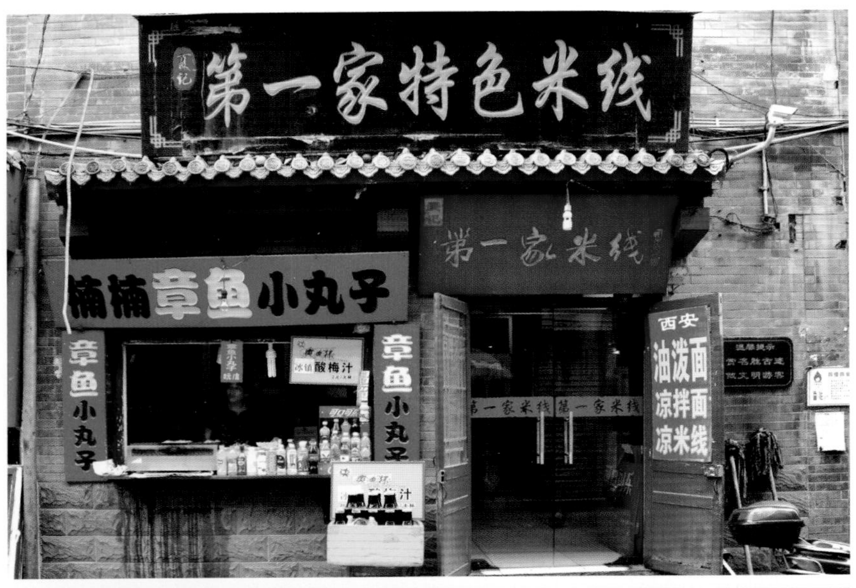

第一家特色米线
第一家米线
楠楠章鱼小丸子
章鱼小丸子
酸梅汁
西安
油泼面
凉拌面
凉米线

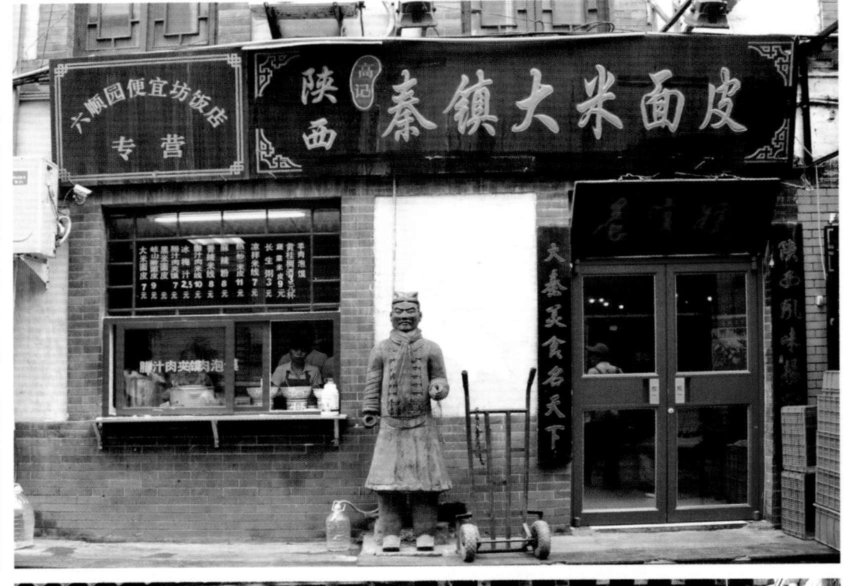

六顺园便宜坊饭店
专营
陕西
秦镇大米面皮
大秦美食名天下

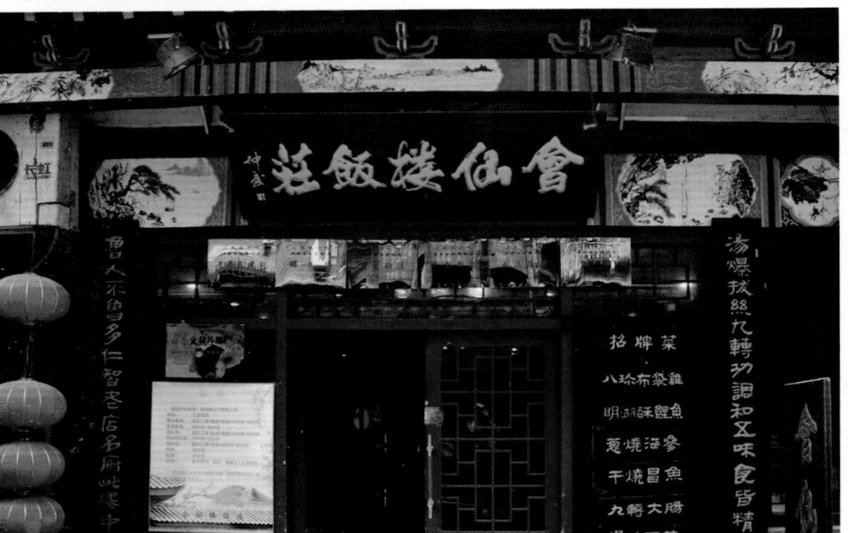

會仙樓飯莊
招牌菜
八珍布袋雞
明湖酥鯉魚
蔥燒海參
干燒昌魚
九轉大腸

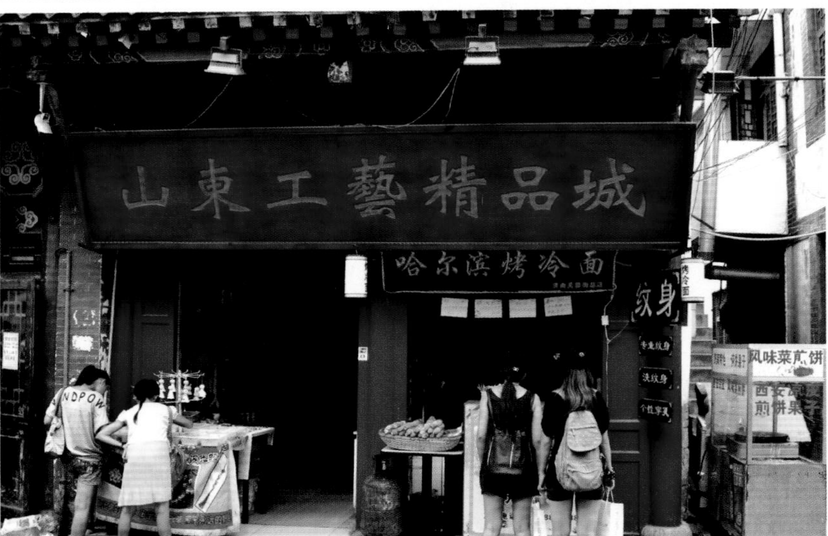

山東工藝精品城
哈尔滨烤冷面
纹身

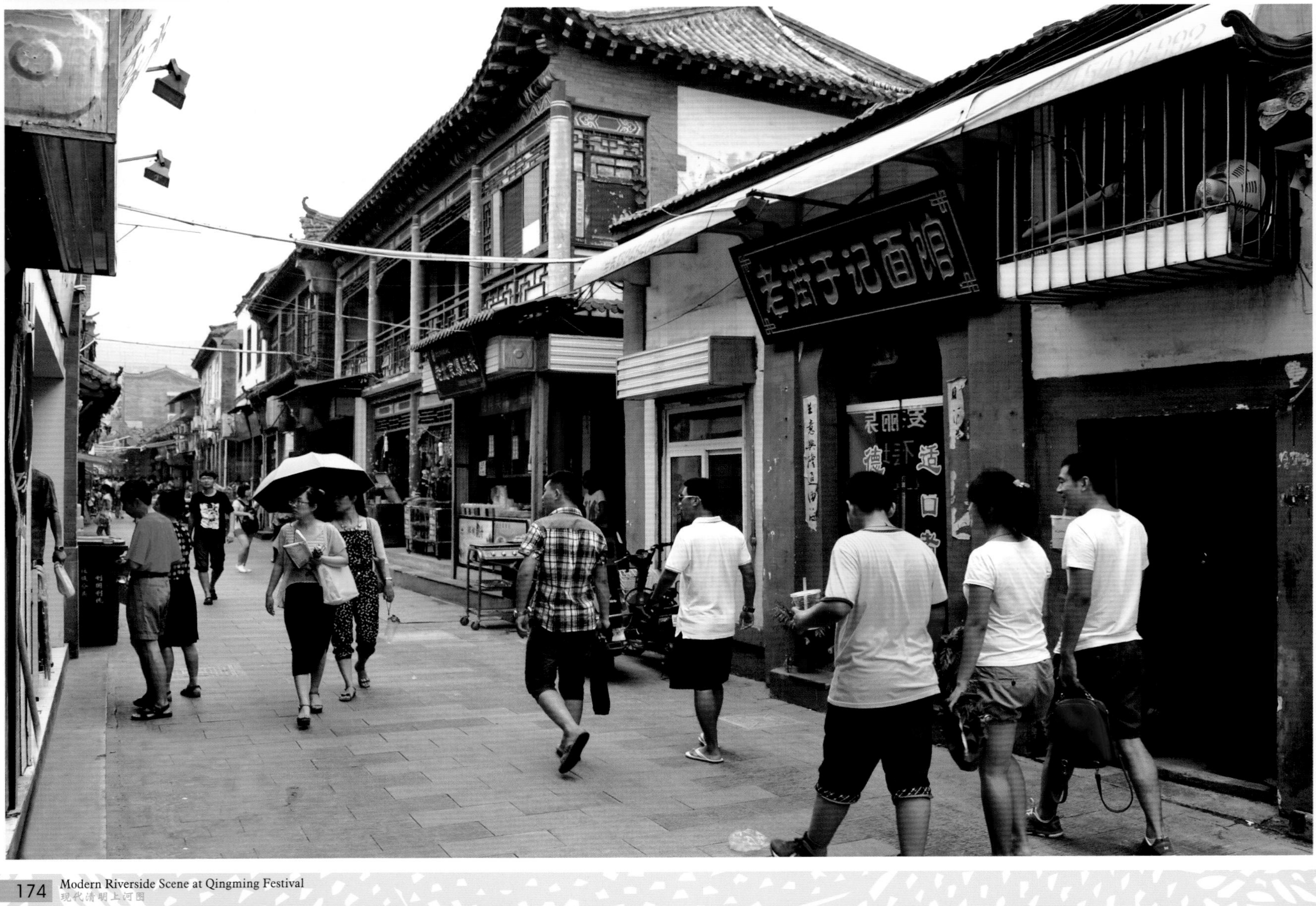

水饺
旅馆

旅馆

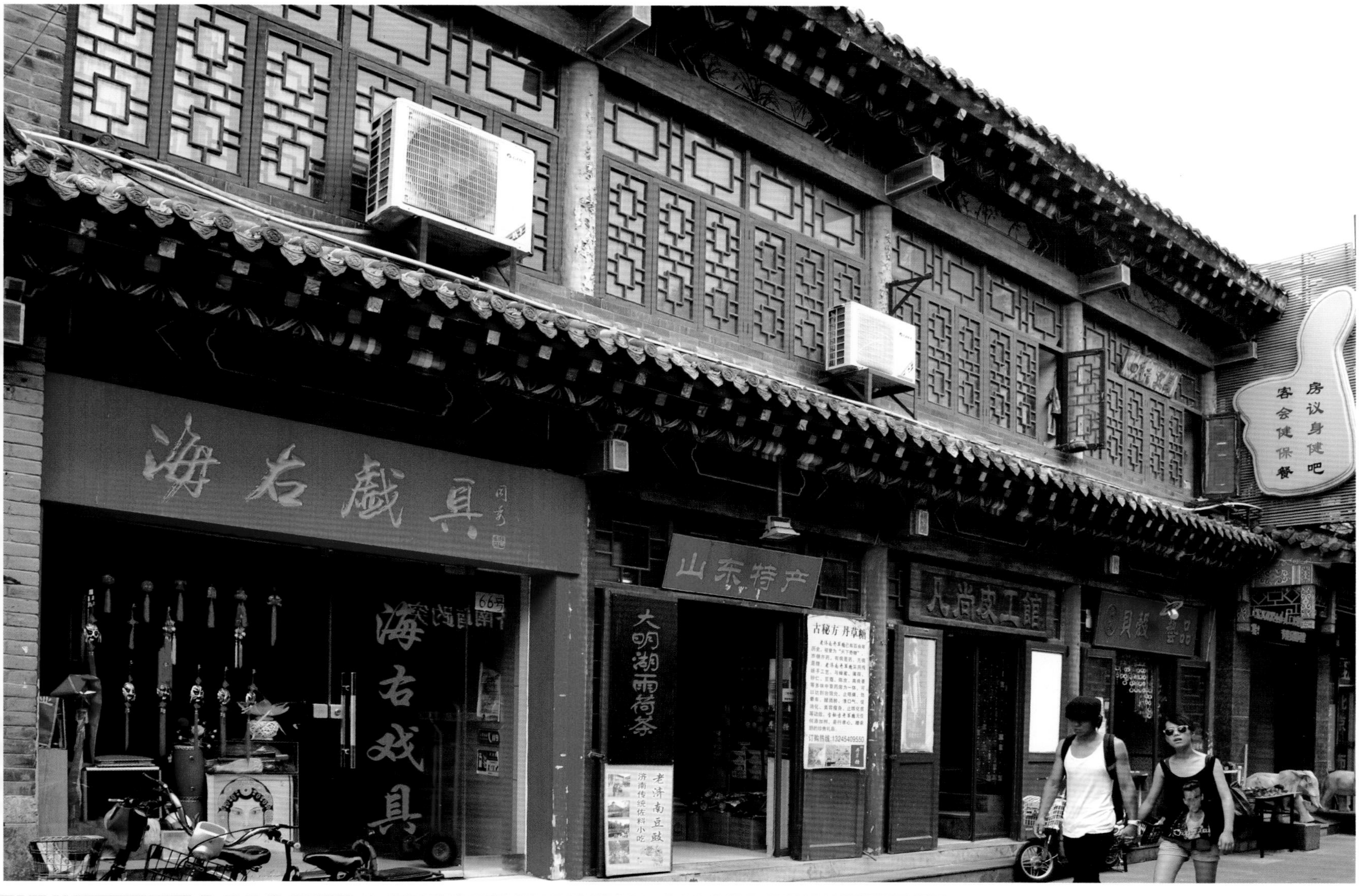

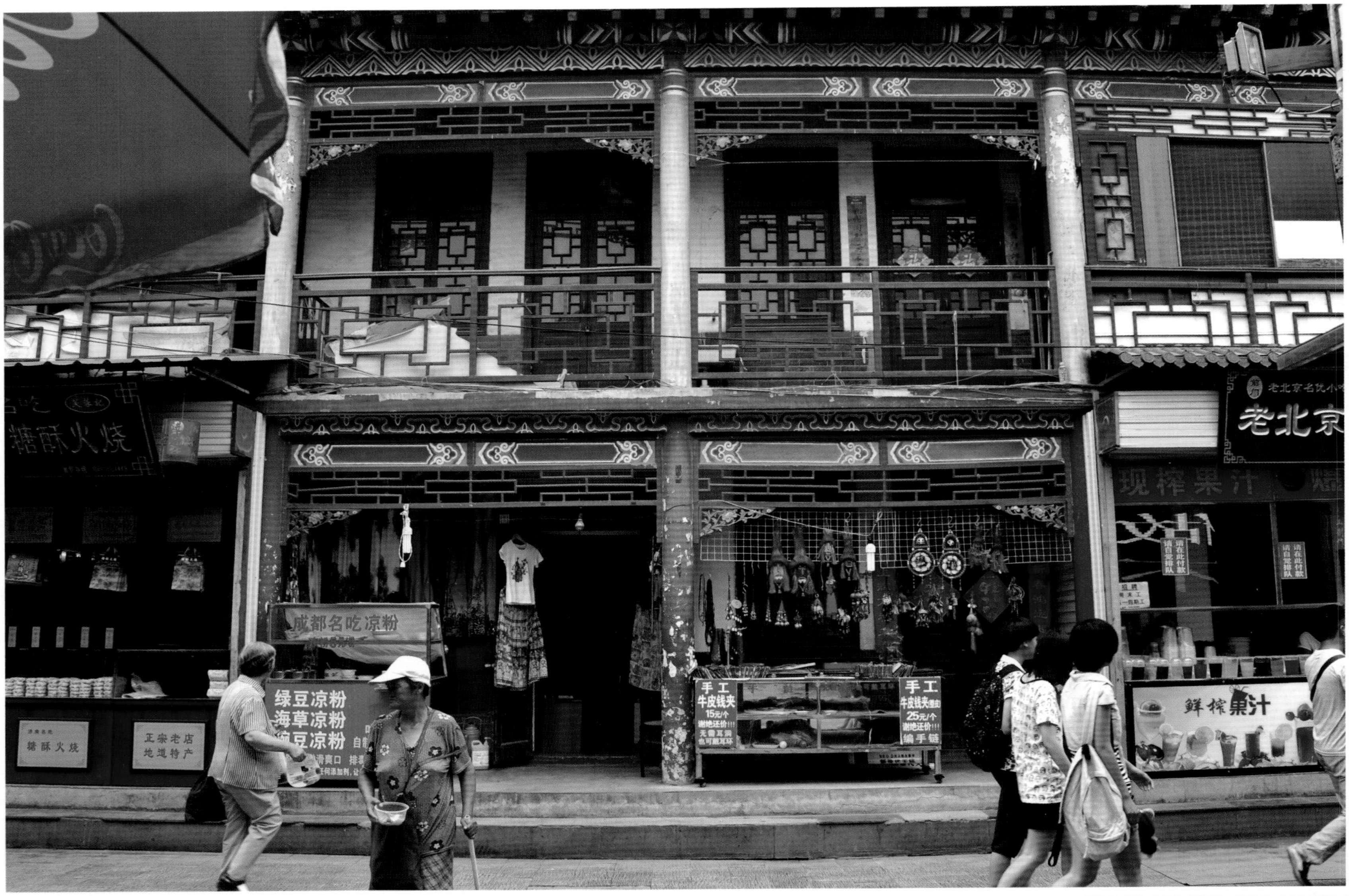

全国连锁 NO.00119
辣尚仙
时尚
焖锅
一锅两吃 辣时尚 仙享受
0531-82892626
好滋味

传统鲁菜
家常炒菜

窗边的小豆豆
休闲吧
芙蓉泉品古城根精气神
阁老爷佑八方客福禄寿

Major Commercial Activities 主要商业业态

After re-positioning, Furong Street, depending on its original natural spring landscape, ancient lanes, ancient residences and cultural relics, was built to integrate multiple commercial forms including tourism, shopping, entertainment and food, and made to be a must-visit scenic spot of appreciating historical culture and daily life of Jinan.

重新定位后的芙蓉街，依托原有的天然泉景、古街巷、古民居和文化古迹，打造成集市民和游客观光、旅游、购物、娱乐、餐饮、休闲等多元化商业的形态，使其成为中外游客领略济南历史文化和感受济南人市井生活的必去景点之一。

Furong Spring 芙蓉泉

Furong Spring is located at No.69, in a rectangular pool built with stones. The pool is ten meters long, five meters wide and three meters deep with a stone bridge cross it and granite handrail at its north.

Furong Spring is the only one of the Jinan Seventy-two Springs located at residence courtyard. At late Ming and early Qing dynasties, there were a small stream between Furong Spring and Pan-chi named "Tiyun Stream", a stone bridge over the stream named "Qingyun Bridge" and a memorial archway written with "rising dragon and soaring phoenix" to imply a prosperous future. But unfortunately, the bridge and memorial archway were ruined during Kangxi period and only left the pool.

Furong Spring is not as unconstrained as Spouting Spring neither as restrained as Pearl Spring. Its beauty of a pretty girl of humble birth got favor from many celebrities. Therefore there were lots of poems written for it from Ming Dynasty to Qing Dynasty. Nowadays, Furong Spring is in the street for sightseeing.

芙蓉泉位于芙蓉街69号居民院前，块石砌长方池，长10米，宽5米，深3米，池中间横跨一石桥，池北侧立花岗石栏杆，原来的池南壁上镌有著名书法家魏启后先生写的"芙蓉泉"碑及手书明代晏璧《芙蓉泉》的诗句，现在已经看不到了。

芙蓉泉是现在位列济南七十二名泉中唯一处于居民院落的泉水，清代郝植恭在《济南七十二泉记》中写道："曰芙蓉，明以艳也。"明末清初时，从芙蓉泉到泮池之间有小溪流，名为"梯云溪"，溪上修有石桥"青云桥"，建有坊额题"腾蛟起凤"的牌坊，寓意前程腾达。康熙年间，青云桥与牌坊具毁，唯余芙蓉泉池。

芙蓉泉不似趵突泉豪放，不似珍珠泉婉转，它以小家碧玉的特色独得众多名士的喜爱。明代永乐年间，诗人晏璧赞曰："朵朵红妆照清水，秋江寂寞起西风。"之后，德王府右史、诗人许邦才在芙蓉泉西建起瞻泰楼。清代康熙年间，诗人田雯题壁瞻泰楼："晴霞飞不断，湖水含泓澄。一丛白菡萏，无数红蜻蜓。"里面的菡萏与芙蓉同是莲花。后还有诗文写瞻泰楼的，有写濯缨湖的，但大抵都是因芙蓉泉的缘故。而今芙蓉泉已经是临街之泉。

Confucious' Temple 府学文庙

Located at the north of Furong Street, the Confucious' Temple was built during North Song Dynasty. It was ruined in a war in Jin Dynasty and rebuilt in the second year of Hongwu period of Ming Dynasty. This temple, which was built following the regulation of Confucian Temple in Qufu, Shandong, has a grand scale with 247 meters long and 66 meters wide at the most. It is facing south with rigorous layout. The building group is built around a south-north axis, orderly, screen wall, Dacheng Gate, Lingxing Gate, Pan-chi and stone arch bridge, screen door, halberd door, Dacheng Hall, Minglun Hall, Huanbi Pavilion and Junjing Pavilion.

府学文庙位于芙蓉街北侧，北宋熙宁年间始建，金朝时毁于战火，明洪武二年重建。文庙仿照山东曲阜孔庙的规制建立，规模宏大，长247米，最宽处66米。建筑群坐北朝南，布局严谨，规模宏大，围绕南北中轴线修建，中轴线上的建筑依次为影壁、大成门、棂星门、泮池和石拱桥、屏门、戟门、大成殿、明伦堂、环碧亭和尊经阁等。

影壁位于文庙之外，向北与大成门对景，其北面正中有圆形砖雕装饰图案，为清代遗物。入庙门往北是棂星门，门前东西两侧设有方亭和圆亭各一座，分别命名为“中矩”“中规”。继续北行，有半圆形泮池，泮池上有石桥。向北屏风的东、西院墙分设“钟英”“毓秀”两座牌楼。向北为戟门，戟门靠近东墙处有一水池，池中有扇面亭一座，名“飞跃亭”。进入戟门，即为文庙的核心建筑大成殿，与东西两庑殿，组成一封闭院落。大成殿建于月台之上，月台前东西两侧各有御碑亭一座。大成殿以北为明伦堂，堂前东西两侧各立两斋，分别为志道斋、据德斋、依仁斋和游艺斋。再向北为尊经阁，阁前设水池，池中有亭名为“环碧亭”。尊经阁以东还有射圃斋，以北有教官宅和儒学门。

Temple of Guan Yu 关帝庙

Temple of Guan Yu was built during Hongwu Period of Ming Dynasty which is over 600 years from now. It is about 600 m^2 with two yards at north and south side. The yard at north side is the principal yard of Temple of Guan Yu. There are two stone tablets inside the temple. One was set up in the fifty-nineth year of Kangxi Reign and written with examination regulations, and the other was set up in the twenty-fifth year of Guangxu Reign to record the rebuilding of the temple.

Around 2010, the government repaired the details of Temple of Guan Yu such as beam columns, doors, carvings and paintings, rebuilt the corridor and pond, and cleaned three ancient wells - Feishuang Spring, Furong Spring, Wuku Spring out. Now the temple of Guanyu is recovered, adding beauty to Furong Street.

关帝庙始建于明代洪武年间，距今有600余年历史，占地面积600平方米左右，分南北两进院子，北院是关帝庙的正院。庙内有康熙五十九年（1720年）《重立考棚碑记》石碑，碑首刻有“共则昭垂”四字，碑文是贡院考试规则。庙内另有光绪二十五年（1899年）《重修芙蓉街北首布政司小街东口路东关帝圣君庙碑记》石碑，乃吉祥号、同祥永、制香楼、信义号、瑞宝楼等二十家商铺为首事，组织两街商家，捐资兴庙，立碑以记。

在2010年前后，政府对关帝庙梁柱、门头、墙面、影壁、砖雕、梁上彩绘等进行修缮，重新修建了碑廊、放生池，将飞霜泉、芙蓉泉和武库泉三处古井挖掘清理出来，现在关帝庙已恢复了历史上的规模和原貌，为芙蓉街增添了新的景观。

The Governor Mansion 巡抚府

Built by Shandong governor Zhou Youde in the fifth year of Kangxi Reign, the Governor Mansion had once received Kangxi Emperor. There used to be a tall and big memorial archway in front of its gate and a screen wall behind the memorial archway. They were demolished in 1930 when Han Fuqu was appointed as the chairman of nationalist party's provincial government.

The Governor Mansion has a grand scale with traditional Chinese style. It was seriously damaged in December, 1937 when Japan attacked Jinan and was reshaped in 1952. In 1979, it was proclaimed as one of the first Key Culture Unites of Jinan.

巡抚府是康熙五年（1666年）山东巡抚周有德在明朝德王府的旧址上用青州衡王府的建材修建而成的，曾接待过康熙皇帝。巡抚部院署占地面积达110多亩，大门前有高大的木质牌坊，牌坊后有影壁墙，后在1930年韩复榘任国民党山东省政府主席时拆除。

大堂面阔五间，进深四间，歇山九脊，翘角飞檐。前为卷棚式，六根大红柱支撑着错落的云头斗拱。檐角脊端，饰以吻兽。1937年12月，日军进攻济南，珍珠泉大院内建筑多被焚毁，巡抚府也遭受严重破坏，1952年被重新修整，恢复原有面貌，1979年被确定为济南市第一批重点文物保护单位。

Tunxi Street, Huangshan

黄山屯溪老街

街区背景与定位

Street Background & Market Positioning

History 历史承袭

Tunxi Street was developed from a ferry which was the joint of Xin'an River, Heng River and Lvshui River in Song Dynasty. In Ming and Qing Dynasties, Huizhou merchants rose sharply. Tunxi by means of its advantage of location became a transportation junction of waterway transport in Huizhou and developed rapidly to be a collecting and distributing center of commodities. The street was considerably influential in Ming Dynasty and became famous in Qing Dynasty. In 1930s and 1940s, for the reason of war and immigration, Tunxi became a significant trading port in south Anhui. After the founding of PRC, Tunxi Street has become one of "China's famous historic cultural streets" to show the culture and charm to the world.

屯溪老街源于宋，由新安江、横江、率水河三江汇流之地的一个水埠码头发展起来。明、清时期，徽商崛起，屯溪凭借地处皖、浙、赣三省交衢，横江、率水回合直通钱塘江的有利条件，成为徽州水路运输的交通枢纽，迅速发展，成为徽州物资集散中心。老街在明朝成为颇有影响的"一邑总市"，清朝时成为闻名遐迩的"茶务都会"。20世纪三四十年代因战乱大量人口迁入，屯溪成为皖南的重点商埠，获有"小上海"的名声。新中国成立后，屯溪老街成为"中国历史文化名街"之一，向世人展示古老的徽州文化，其魅力倾倒了国内外的旅游者和影视制作者，老街成为天然的摄影棚。

Location 区位特征

Tunxi Street is located at the center area of Tunxi District, Huangshan City. It extends from Zhenhai Bridge which is a stone arch bridge built in Ming Dynasty to memorial archway at the east. It is over 1,000 meters long and 4.8 to 7 meters wide.

屯溪老街坐落于黄山市屯溪区中心地段，西起于明朝时期建立的石拱大桥镇海桥，东至牌坊碑记，街道全长1 000余米，宽4.8~7米。街区历史悠久，商业繁华，每年吸引近600万的国内外游客观光旅游。

Market Positioning 市场定位

In 2011, Tunxi Street was comprehensively remolded according to five functional positionings - culture exhibition, business communication, leisure entertainment, tourist reception and featured snack, to make the street a top grade historic cultural street assembling tourism, business, culture, restaurant and entertainment.

2011年，屯溪老街实施全面改造，提升项目建设，按照"文化展示、商贸流通、休闲娱乐、旅游接待、特色小吃"五大功能定位，重点抓正街整治、小巷开发，全面提升屯溪老街的文化、旅游品位，丰富老街经营业态，努力将屯溪老街打造成一流的历史文化街区，使其由原来单纯的商业购物街转向集旅游、商贸、文化、餐饮、娱乐于一体的街区。

Street Planning 街区规划

The planning of Tunxi Street adopts policy of "protection, repair and renovation" and obeys the reality of history, completeness of history appearance and continuity of life. The planning protects the interior of some featured tradition buildings as well as fasades in the street. Moreover, the planning puts emphasis on protecting traditional space layout.

The protection section of Tunxi Street is divided into protection zone, control zone and coordination zone which are protected and renovated by different measures. Historic buildings are classified into different grades, and original façades, structure systems, plan layouts and interior decoration of significant buildings are preserved.

屯溪老街街区规划采取"保护、整治、更新"的方针，秉承保护老街历史载体的真实性、历史风貌的完整性和生活的延续性的原则。规划从保护老街街面建筑外观，深入到一些有特色的传统建筑内部原貌的保护；重点保护老街核心区，深入到老街背后两侧街区传统空间格局、肌理的保护。

规划将屯溪老街保护地段划定为保护区、建设控制区和环境协调区三个层次，采取不同的措施予以保护和整治；划分建筑等级，保护历史建筑，重要建筑原有的立面、结构体系、平面布局和内部装饰一律不改变。

Street Design Features 街区设计特色

The buildings in Tunxi Street are of traditional Huizhou ancient building style. Most of the shops along the street are of two floors with shops on the first floor and dwellings on the second floor. The layout and architectural form of the building group in the old street have distinct features of Huizhou style architecture which is quietly elegant and antique. The structure and design of the buildings are vey unique. The shops on the street are mostly single rooms and are partitioned to each other by fire walls.

Tunxi Street is a display window for great and profound Huizhou culture. There are Huizhou style architectural culture featuring fire walls and carvings, Xin'an medicine culture represented by Tongderen drugstore, Xin'an painting and calligraphy culture represented by horizontal inscribed boards, couplets, Huizhou cuisine culture featuring Laojie Diyilou, and the "scholar's four jewels" culture, Huizhou tea culture and so on. All these present the connotation of Huizhou culture truly.

屯溪老街建筑具有传统的徽州古建筑风格，沿街店铺大都二层，下店上房，前店后坊形制建筑体量有十多万平方米。老街的建筑平面既有沿街敞开式，又有内天井式，建筑结构注重进深，所谓的“前面通街、后面通河”往往是大店铺的格局。老街的建筑群的规划布局、建筑形式具有鲜明的徽派建筑特色，建筑体量大小相间，色彩淡雅、古朴。老街建筑的结构和款式独具特色，小青瓦、白粉墙、马头墙，古色古香。街中店面多为单开间，店铺之间有马头墙封护相隔，屋面盖小青瓦。底层门面采用木排门，便于营业。这种入内深邃、连续几进的房屋结构形成了屯溪老街前店后坊、前店后仓、前店后居或楼下店楼上居的经营、生活方式。

屯溪老街是博大精深的徽州文化的集中展示窗口，形成了以粉墙黛瓦、马头墙和砖雕、石雕、木雕为主要特征的徽派建筑文化，以同德仁药店为代表的新安医学文化，以书画、匾额、楹联为代表的新安书画文化，以老街第一楼、老徽馆为代表的徽菜文化，以歙砚徽墨为代表的文房四宝文化，以三味茶馆等为代表的徽州茶文化以及以馆藏器物和工艺品为代表的民间器物文化等等，真正体现了徽州文化的内涵。

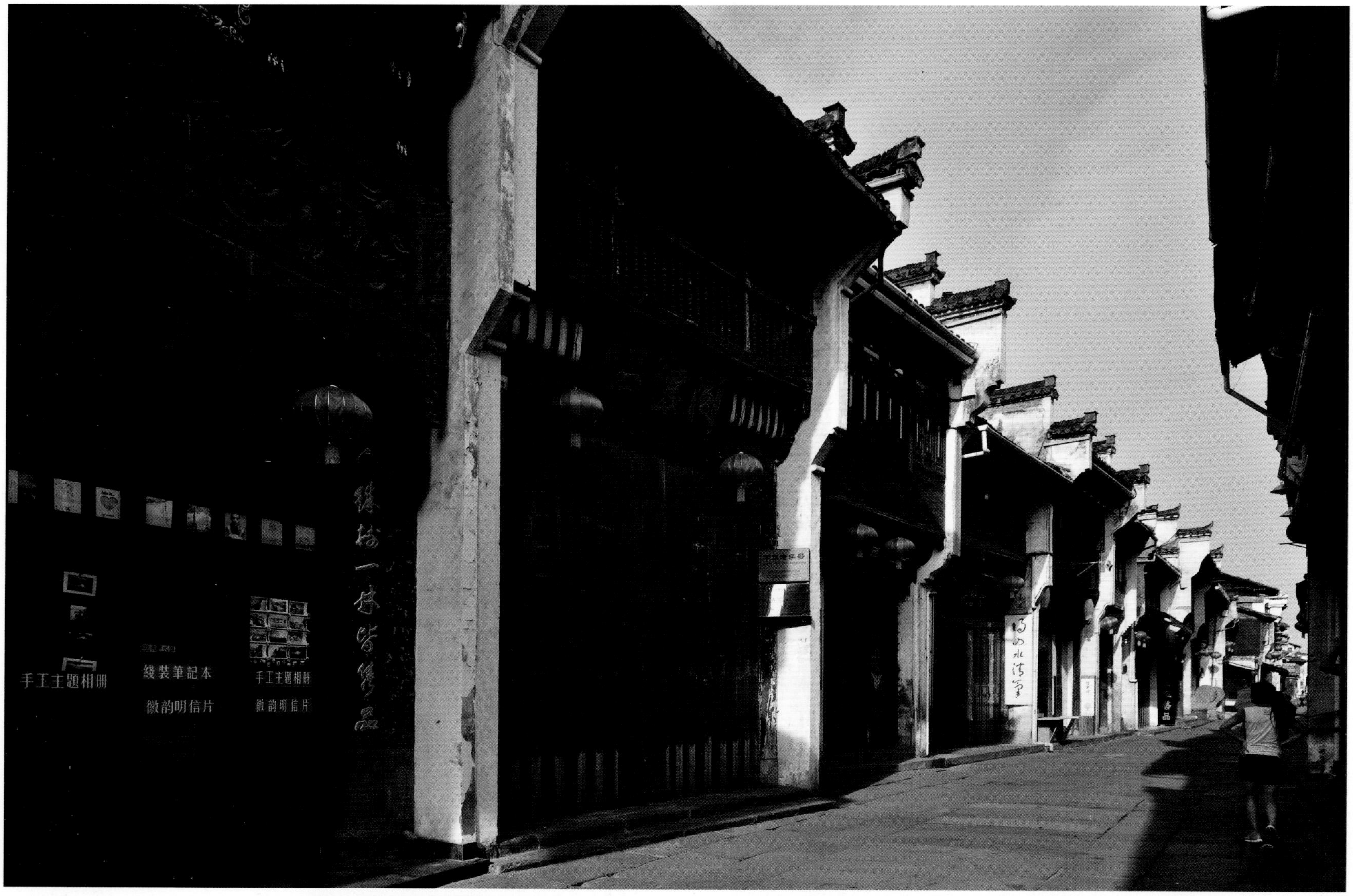

缅甸天然翡翠专营
天然 翡翠
翡翠特
2.8

藏
寶
西
纳

Major Commercial Activities 主要商业业态

There are over 300 shops including more than 60 time honored shops in Tunxi Street. Tunxi Street is composed of Zheng Street and No.1, No.2 and No.3 Roads. The Zheng Street is the main composition for sightseeing, and the other roads feature curios, leisure and snacks. Therefore a commercial layout of "the vertical street as the major and horizontal roads as the supplement" is formed.

No.1 Road features curio with more than 30 shops running curios.
No.2 Road featuring leisure has bars for tea-tasting and rest.
No.3 Road featuring snacks offers many local snacks such as Huizhou wontons, sesame paste and so on.

屯溪老街共有店铺300余家，其中有60多家老字号店铺。屯溪老街包括正街和一马路、二马路、三马路，游客以游玩正街为主，经过业态规划引导和政府的精心组织实施后，老街一马路、二马路、三马路形成了以古玩、休闲、小吃为主的特色街区，整个街区形成了“以纵为主，以横辅纵”的商业格局。

一马路是古玩一条街，以经营古玩的店铺为主，现有30余家经营古玩的店面。

二马路是休闲一条街，拥有德阳楼、过客、阁楼等酒吧、茶吧，是游客品茶休憩的好地方。

三马路是小吃一条街，街中有地道的徽州馄饨、手抓饼、芝麻糊等特色小吃，也有梅干菜扣肉、臭鳜鱼等正宗徽菜。

Operation Measure 运营措施

1. To enrich business activities in Tunxi Street

According to its five functional positionings and on the basis of improving three featured streets, the old street part of Binjiang Road is reconstructed to build bars, thus, enriching the business activities and enhancing popularity.

2. To introduce marketable operation mode.

The marketable operation mode is to manage the old street as an "enterprise" and sell tourism as a product by introducing the concept of enterprise management, meanwhile strengthening the cooperation with movie and TV medias to advertise and promote Tunxi Street.

3. To innovate manage system of the street

The management should be strengthened. All the shops in the street should be honest and be allowed by the managing office of the street. All the reconstruction and repairing of the houses in the street should be agreed by the office and constructed by appointed unit. All motor vehicles are strictly forbidden.

4. To explore the system of "supporting the street by the street"

Take reference of the successful practice of Old Town of Lijiang and explore a new way to protect and develop the street. The protection problem of Tunxi Street could be solved by collecting protection fee from visitors.

1.丰富屯溪老街业态

按照"文化展示、商贸流通、休闲娱乐、旅游接待、特色小吃"五个方面的功能定位，在完善三条特色街打造的基础上，改造滨江路老街段，沿江建设具有浓郁徽州特色的茶吧、酒吧、咖啡吧和能提高游客参与度的休闲场所，丰富业态，提升人气。

2.引入市场化运作模式

引入企业管理的概念，将老街作为"企业"，将老街旅游作为"产品"营销，通过招商引资，将老街推向市场，力争引入实力雄厚的公司和更多社会资本进入。同时，加强与影视传媒合作开发制作电影、电视片，利用传媒，全面宣传推介屯溪老街，使老街更具魅力，更具看点。

3.创新街区管理机制

加强前置管理，所有进入老街经营的店铺，必须承诺"门前三包"、诚信经营，经营许可必须由老街管理办签字；所有房屋的改建修缮必须由管理办签字同意，按规划要求，由指定的施工单位施工；所有机动车辆由交警实行严管严罚。

4.探索"以街养街"机制

借鉴丽江古城的成功做法，探索街区保护利用的新路子，通过全市旅行社向游客收取古街保护费，以解决屯溪老街保护问题。

Food & Beverage 餐饮类

Laojie Diyilou

Laojie Diyilou located at the east entrance of Tunxi Street is a famous local Huizhou cuisine enterprise. From the lobby to every box, the whole building reveals on old charm of Huizhou style. All the dishes here are pure traditional Huizhou cuisines which are various and delicious.

老街第一楼

老街第一楼位于屯溪老街的东入口，是当地一家知名的徽菜餐饮企业。老街第一楼从大厅到包厢处处都体现出徽风古韵，各式各样的徽派建筑元素精心点缀其中，每一处都有独自的主题。楼中的菜式是纯正的传统徽菜，如“徽州毛豆腐”“一楼如意鸡”“徽州问政笋”等，品种繁多，口味地道。

Xieyu Tea House

Xieyu Tea House was founded in 1875 by Xie Zheng'an, a tea making professor in the late Qing Dynasty who was famous for creating "Huangshan Maofeng" tea. Xieyu Tea House engages in developing and selling famous teas. In the late 19 century and early 20 century, Xieyu Tea House covered its business not only in China, but also in the Far East and Western Europe.

After the reform and opening-up, Xie Yiping, the great-great-grandson of Xie Zheng'an rebuilt Xieyu Tea House. In 2012 the tea house was successfully turned to Huangshan Xieyu Tea Co., Ltd. and was authorized as "China Time-honored Brand" by National Commerce Department.

谢裕大茶行

谢裕大茶行创办于1875年，缔造者谢正安——晚清时期的制茶专家，因其创制的“黄山毛峰”而永载史册。谢裕大茶行致力于研发、销售原产地名茶，早在19世纪末20世纪初，谢裕大茶行就遍布神州各地，并将茶叶远销远东和西欧，有“名震欧洲四五载”之美誉。张之洞曾为其题下“诚招天下客，誉满谢公楼”的赞词。后又有新安画派大师黄宾虹赞谢裕大茶行为“黄山毛峰第一家”。

改革开放后，谢正安五世玄孙谢一平重建谢裕大茶行，2010年成功改制为黄山谢裕大茶叶股份有限公司，并获得国家商务部“中华老字号”认定。

Yipin'ge

Yipin'ge is located at No.251 Tunxi Street with a total operating area of about 200 m². It mainly sells various famous teas in Huangshan. The façade of its building inherits Huizhou architecture style and its interior environment is elegant. It is a good place for tea-tasting and experiencing tea culture.

一品阁

一品阁位于屯溪老街251号，总营业面积近200平方米，主营曹溪牌黄山毛峰、猴坑牌太平猴魁、祁门红茶等各种黄山名茶。一品阁建筑外形继承了徽派建筑风格，室内环境优雅，茶器古典，是品茶、体验茶文化的好场所。自2002年以来，一品阁获得了“诚信商店”“消费者放心店”等殊荣。

Fangsheng Tea

Huangshan Fangsheng Tea Co., Ltd. is named after "Wang Fangsheng", its fifth generation successor. Its subordinates include several ecological tea garden basements and tea factories located at natural forest area in Huangshan City. Its producing process has gotten national patent. Its products include hundred kinds of craft molding tea and are sold in more than eighty countries and regions.

芳生茶业

黄山芳生茶业有限公司以其第五代茶人"汪芳生"命名，公司名称由中央民盟名誉主席、国际茶道联合会副会长谈家桢教授题写。其下属有多个生态茶园基地和茶厂，分布在黄山市天然林区的千米高山峡谷之中，以历来不喷雾化化学农药、化肥的正宗生态茶叶为原料，生产加工技术荣获国家专利，有"锦上添花""神龙茶""双喜临门"等数百个品种的工艺造型茶，茶叶色艳、毫显、香高、汤清、味甜、形美六绝，行销于80多个国家和地区。

Mingxiang Tea House

Mingxiang Tea House is an outlet store of Huangshan Huizhen Food Co., Ltd. Huangshan Huizhen Food Co., Ltd. was founded in 2003 and engages in producing and selling almost one hundred kinds of subsidiary agricultural products including tea, chrysanthemum, dry goods, crunchy candy and so on.

茗香茶庄

茗香茶庄是黄山市徽珍食品有限公司旗下的直销店，黄山市徽珍食品有限公司创建于2003年，主要从事生产经营茶叶、菊花、山珍干货、茶食品、酥糖糕点等上百种农副产品。茗香茶庄主营黄山毛尖、太平猴魁、忆丝清徽菊等品种的茶叶。

Yixinxiang

Yixinxiang Tea was founded in Guangxu Period by Sun Qiben and was developed by his descendants. Now it inherits its tradition and meanwhile innovats boldly to integrate a profound culture into the new era. From protection of tea garden to package, it expresses the nature, purity, freshness and health of tea from various aspects and elaborately produces every kind of art tea.

怡新祥

怡新祥茶号始创于光绪年间，创始人孙启本，后世子孙迁号徽州屯溪观音山，将其所产茶叶销至国外地区。现在的怡新祥人秉承传统，大胆创新，将一份厚朴的文化融入新的时代视角，从茶园保护、鲜叶采制到包装品饮，多个角度阐释茶的天然、纯净、质朴、清新、健康、和谐的特质，精心打造每一款纯天然艺术茶。

Craft 工艺类

Maohuai

Maohuai is the largest shop selling antique arts and crafts in Tunxi Street. It is also the most completely preserved Huizhou merchants' building of "front shop and back dwelling"in the street.

茂槐

茂槐是屯溪老街最具规模的经营古旧工艺美术品的店铺，也是老街保留最完整的“前店后居”式徽商建筑。店内有精美的石雕与木雕。

Bamboo Art House

Bamboo Art House is only 50 square meters. It mainly sells products from the eight world expo licensed merchandise production enterprises. Most of products are bamboo carvings of exquisite shapes and present the strong Huizhou cultural features.

竹艺轩

竹艺轩面积不大，仅有50平方米，里面主营黄山市8家世博会特许商品生产企业的产品，多是造型优美、制作精美的竹雕工艺品，体现出浓浓的徽州文化特色。

300-Ink-Stone Shop

300-ink-stone Shop is located at the most prosperous part of the street and specializes in producing and selling She ink stone. Its products are made of excellent materials by fine workmanship, not only practical, but also admirable and worthy of collection.

300-ink-stone Shop once made ink stone for celebrities such as Wu Zuoren, Shen Peng, Xie Yun and Shao Yifu, Princess Sirindhorn. It was praised as "top producer of She ink stone" by Wu Zuoren, a Chinese painting master and a former chairman of Chinese Artists Association. Jiang Zemin, former state president once encouraged 300-ink-stone Shop to inherit the culture of ink stone generation by generation and make it forever.

三百砚斋

三百砚斋位于老街最繁华的地段，店名为我国著名书法家沈鹏先生所题。三百砚斋专事歙砚的生产与销售。所制作的歙砚选材精良，创意精巧，雕刻精工，其作品融诗书画印砚于一炉，集实用、观赏、收藏于一体。

三百砚斋曾为当代著名书画家吴作人、沈鹏、谢云及国内外友人邵逸夫、诗琳通公主等制砚。其制作的国礼砚"黄山胜迹印痕"砚，1997年由李鹏总理送给日本天皇明仁陛下。它被已逝中国绘画大师、前美协主席吴作人题书赞誉为"歙砚第一家"！前任国家主席江泽民视察三百砚斋时勉励道："如此灿烂的文化，如此博大精深的文化，一定要世世代代、子子孙孙传下去，让它永远立于世界文化之林。"

Yang Wen Pen Studio

Yang Wen Pen Studio is named after its owner Yang Wen who is the director of Huangshan City Hui Pen Craft Research Institution and the inheritor of intangible cultural heritage. Brush pens made by his ancestors were tributes. Vice-chairman of Chinese Artists Association Du Ziling once greatly praised "Ziling Magic Pen" made by Yang Wen. Now there are totally over 20 researchers in Yang Wen Pen Studio and Hui Pen Craft Research Institution with an annual output of more than one hundred thousand pens.

杨文笔庄

杨文笔庄因店主杨文而得名，杨文是黄山市徽笔工艺研究所所长，非物质文化遗产传承人。其祖辈所制的毛笔曾为贡品。中国美术家协会副主席杜滋龄先生曾对杨文研制的“滋龄妙笔”赞不绝口，不但为其题写“杨文笔庄”，还用此笔画了一幅《藏族风情》赠予杨文。现在“杨文笔庄”与“徽笔工艺研究所”内共有制笔、研究人员20多人，年产徽笔十余万支。

Riyuelou

Yiduoyiguo is in the Riyuelou and sells unique notebooks. Its signboard is black base with golden words, simple and traditional. This shop is small with ranks of bookshelves displaying various kinds of unique notebooks. The interior walls of the shop are full of tags which are left by the customers or visitors.

日月楼

一朵一果是一家经营个性笔记本的店铺，坐落在老街的日月楼中。店铺的招牌是黑底金字，简单而传统。店铺不大，里面一排排的书架上陈列着多种多样的个性笔记本，店的内墙满是贴纸，都是来此购物或游览的人留下的。

Traditional Chinese Medicine 中药类

Tongderen

Tongderen was founded in 1863 by Chen Dezong and Shao Yuanren. The shop is named after the two founders and means cooperation and justice. Tongderen is an herbal medicine shop selling traditional Chinese medicine. The shop building is of brick and wood structure. Through over one hundred years ups and downs, it is still the biggest time-honored medicine shop in the street.

同德仁

同德仁创办于清朝同治二年（1863年），由程德宗、邵远仁合伙开办，店名便从他二人名字中各取一字，寓意“同心同德，办事仁义”。同德仁是一家经营中药的药铺，店面为两层的砖木混合结构，历经百余年的风风雨雨，仍然是老街上规模最大的老字号药店。

Tianlaigu Museum 天籁谷博物馆

Tianlaigu Museum is divided into a shop in the front and a museum at the back. The shop mainly sells potteries, ancient jades and paintings. The museum mainly exhibits weapons, seals and jade articles.

天籁谷博物馆前店后馆，前面店铺以陶器、古玉和馆主自己画的油画为主，油画主要是黄山的景色，后面的博物馆主要展示兵器、印章、玉器等物品。

Tunxi Museum 屯溪博物馆

There are three special exhibitions in the museum – Huizhou brick carving art exhibition, Huizhou character image exhibition and Ming and Qing dynasties furniture exhibition. It mainly shows brick carvings, paintings and ancient furniture to visitors.

屯溪博物馆内设有三大专题展：徽州砖雕艺术展、徽州人物容像展、明清家具展。主要向游客们展示徽州的砖雕艺术、绘画艺术、古代家具艺术以及从中延伸出的古徽州风土人情、纲常礼教等。

Wancuilou 万粹楼

Wancuilou is the first ancient building museum in the country. It was opened in April, 1999. The building is 24 meters tall of four floors and over 500 square meters. Its interior and exterior decorations and landscapes are of traditional Chinese style which is antique and exquisite.

Its second floor is an exhibition hall displaying cultural relics of past dynasties; the third floor with front hall and back lane is a residential living room of Hui style; the fourth floor is a roof garden of Hui Garden style.

万粹楼是我国第一家古建筑博物馆，1999年4月对外开放，由全国政协副主席叶选平题写"万粹楼"楼名。万粹楼楼高24米，上下4层，占地面积500多平方米，雕梁画柱，飞檐翘角，集木雕、砖雕、石雕为一体，古料新建，匠心独运。楼内飞梁红柱，宽大轩敞，一池清水，石雕小桥，轩顶挂落，精雕细刻。古董家具摆设堂前。

楼内二楼为展厅，陈列历代文物，名家瓷板艺术；三楼前厅后巷，弄外回廊，古色大画，一派徽式居民斋厅；四楼是屋顶花园，青瓦白墙，花坛盆景，是徽式的园林风格。

Huangshan Land International Center

黄山置地国际中心

街区背景与定位 Street Background & Market Positioning

History 历史承袭

Liyang was established as a county subordinating to Xindu Prefecture in Jian'an thirteenth year of the Eastern Han Dynasty. Because of its location at the joint of two big rivers and three provinces, it was a commercial center of Anhui, Zhejiang and Jiangxi, and also a significant port town of Xin'an River. However, with the change of times, it lost its property.

黎阳老街素有"唐宋之黎阳，明清之屯溪"的美誉。东汉建安十三年（公元208年）设犁阳县（后改为黎阳），属新都郡；西晋太康年间(公元280−289年)，新都郡改新安郡。至南北朝陈文帝天嘉三年(公元562年)，黎阳县两度并入海宁县；隋文帝开皇十八年(公元598年)定名休宁后，黎阳属休宁县的一个乡，屯溪即是黎阳乡的一个都（古时村庄别名），又称休宁县黎阳十六都。"两江交汇，三省通衢"的地理位置，使黎阳成为皖、浙、赣边陲商业中心和新安江的码头重镇。然而，随着时代的变迁，黎阳古镇失去了往昔的繁华，街巷破败不堪，房屋损毁严重。

Location 区位特征

This project is located at Liyang Town, Tunxi District, Huangshan City, Anhui Province. It situates at the joint of Xin'an River, Heng River and Lvshui River, and the traffic node of Airport Avenue, Hetonghuang Freeway and Huihang Freeway, therefore, it enjoys an advantageous location.

项目位于安徽省黄山市屯溪区东南角黎阳镇内，基地西侧、北侧为占川河，南侧濒临率水河，处于新安江、横江、率水三江交汇处，临机场大道、合铜黄高速、徽杭高速等重要交通节点，地理位置优越。

Market Positioning 市场定位

It gathers historic cultural area (Liyang Ancient Town), Huizhou cultural street, international hotel and holiday apartments, international tourism reception center, and others to create international tourism service complex of tourism, leisure, holiday and dwelling.

项目聚合历史文化街区（黎阳古镇）、徽州文化风情街（徽文化水街）、国际酒店及度假公寓群、国际旅游接待中心、徽派体验式住宅、高档住宅多种物业，打造一个集旅游、休闲、度假、居住为一体的国际级旅游服务综合体。

规划设计特色
Planning & Design Features

Street Planning 街区规划

It was mainly composed of Riverside Bar Street, Liyang Ancient Street Culture Area and Featured Food Area.

Riverside Bar Street: it makes use of the advantageous natural resources of the three rivers estuary. The bars, cafes and teahouses along the river are designed in arch building style. The urban leisure environment is created by the style and taste of bar culture.

Featured Food Area: various local food and snacks attract customers. And details of buildings add to the cultural features and experience of this area.

Liyang Ancient Street: on the basis of keeping the original texture of old building, museums, galleries and time-honored shops are arranged to complete the sightseeing and tourism environment. Cultural landscapes are added to increase its enjoyment and to more sufficiently show the Huizhou culture.

项目由滨江酒吧休闲街区、黎阳老街文化街区、特色美食街区等部分构成。

滨江旅游休闲带：利用三江口优越的自然资源，沿江采用骑楼式布局设计酒吧、咖啡馆、茶楼，利用酒吧文化所代表的格调、品位，营造出黄山国际化旅游城市休闲氛围。

风味特色餐饮街：街中设民间各类特色餐饮、风味小吃，发挥了汇集人流作用；通过设计剧院等相关公共建筑，以及建筑细节、小品构件渲染，增强街区的民俗文化特色和体验性。

黎阳老街：在保留原建筑街巷肌理上，布局博物馆、国医堂、国画堂、徽州老字号商行、旅游纪念品店、土特产店等商业，充分打造贾宅、石宅的观光旅游氛围。沿街主轴线点缀徽州文化小品，增强趣味性，充分展示徽文化，再现悠久的徽派建筑风格。

Street Design Features 街区设计特色

The design purpose is "protection, transplanting, innovation". In order to preserve the memory of Liyang Ancient Street, the general design keeps the winding stone street texture.

The design organically combines new buildings with the ancient street. It continues the feature of exterior space. Street space, node space and square space are organically integrated to make an effect of changing scenes depending on how visitors move about. Stone pavement, doors at both sides, and exquisite brick carves on doors enrich the texture of the whole space.

The plane design of a single building, on the basis of keeping the traditional interior space layout of Huizhou residences, is improved in architectural scale and living facilities according to modern life style. And by adopting traditional architectural features of Huizhou, burgeoning tourism experience buildings and buildings for both business and residence are formed.

Hui-style building is not only an important part of Huizhou geographical environment, but also the material carrier of Huizhou culture. This project is based on Hui-style architecture aesthetic, and references to essence of Huizhou typical architecture form. According to different locations and function positionings, buildings are designed as traditional style and modern style, yet keep the essence of Hui-style.

Meanwhile, on the basis of keeping original street texture and repairing part Ming Dynasty of old residences, traditional architectural decorations are applied in forms of courtyard and shop and featuring Huizhou architectural style of Ming Dynasty. On the basis of keeping the traditional architectural features, the entertainment area at the south bank is planted with modern architectural language, for example, steel frame glass buildings, showing energy and passion of the street.

The general design pays a great attention to the protection and exploration of culture of Liyang Ancient Street. It tries to reproduce the prosperity of Liyang Street. Therefore, visitors can learn the profound history of the street while feeling the modern urban features.

在设计中，以“保护、移植、创新”为宗旨。为了保存黎阳老街的记忆，老街以北以贾宅、石宅及老建筑为界，以南保留破损房屋基底边际线，整体设计保持了原青石板蜿蜒曲折的街道肌理。

设计加入徽州水系特点，将新建筑与老街有机结合。延续并继承了徽州外部空间的特点，将建筑以线状的街巷空间、点状的结点空间（如街巷交汇点，院落空间）、面状的广场空间及线面结合的水口空间有机结合，达到步移景异。铺地的青石板，两侧的门坊、门罩上精致的砖雕，与天际线交接的马头墙檐部，甚至粉墙上留下的斑驳印记，都丰富了空间的肌理。

建筑单体平面设计在保持传统徽州民宅内部空间“天井明堂”“四水归堂”“纵深序列”等基本构成方式的基础上，依据现代生活方式在建筑尺度、生活设施等方面加以改进，采用以传统徽州建筑特色的窗棂、雀替、美人靠、木雕、石雕、仿古家具等形成新兴的旅游体验式建筑及商住两用的建筑。

徽派建筑是徽州地域环境的重要组成部分，而且是徽州文化的物质载体，“粉墙青瓦马头墙”构建了古徽派建筑的审美主题。项目在建筑形态上立足于传承徽派建筑审美意向，取意于徽州典型建筑形式的精髓，根据不同区域位置及功能的定位，建筑设计为传统建筑风格与现代新徽州风格，总体保持徽风徽韵的精粹。

同时，在保持原老街机理及保护、修缮部分老民居的基础上，以院落及少量店铺式艺坊的建筑形式，以明代徽州建筑风格为特色，采用传统的建筑装饰符号，如商字门、石雕漏窗、抱鼓石、大面积的白墙与少量的木质店面相对应。南岸的娱乐休闲区在保持具有传统建筑特征的基础上，植入现代建筑语言如纯钢架玻璃建筑等，使街区呈现出新的活力与激情。

整体设计重视黎阳老街文化原生态的保护与发掘，力求重现黎阳老街旧日的繁荣与辉煌，让游客走进黎阳老街既能体味千年古镇的厚重历史，又能感受到现代都市的时代特征。

漫咖啡酒吧
M COFFEE &

咖啡酒吧

LINONG LANE

WELCOME
MON-FRI
AM 10:00 ~ PM 10:00
SAT-SUN
AM 10:00 ~ PM 11:00

Major Commercial Activities 主要商业业态

Food & beverage: big Chinese restaurants, theme restaurants, cultural restaurants, Western restaurants and featured snacks.
Leisure: music clubs, fashionable bars, cafes, amusement bars.
Culture: Huizhou culture experience, culture salon, specialties, curios, accommodation.
Wedding celebration: wedding etiquette, wedding photography, flowers and gifts.

Hotels are in modern architecture form with partial Hui-style elements. Hotels of different levels are standing intensively to form a well-proportioned symbol building group.

In the street area, customers can appreciate traditional arts, such as three carves of Huizhou – wood carves, stone carves and brick carves, the scholar's four jewels, poems, dramas and so on. Meanwhile, with modern support facilities, concept of creative fashion is integrated into Hui-style culture including food, scenery, leisure, entertainment and culture to provide customers with comfortable experience and unique cultural enjoyment.

餐饮类：大型中餐、主题餐饮、文化餐厅、西餐厅，特色小吃。
休闲类：音乐会所、时尚酒吧、咖啡厅、休闲吧。
文化类：徽文化体验、文化沙龙、特产、古玩、住宿类。
婚庆类：婚庆礼仪、婚纱摄影、鲜花礼品。

酒店以现代建筑形式为主，局部注入徽派元素。五星级酒店、四星级酒店、三星级酒店、特色酒店、各种等级的公寓式酒店等集中排布，形成错落有致的标志性建筑，与徽商中心形成颇具规模与竞争力的酒店群。

在街区内，可以品味徽州三雕——木雕、石雕、砖雕，笔墨纸砚文房四宝，诗歌、戏剧、新安画派等传统艺术；同时，用现代化的高端配套服务作为支撑，将创意时尚的理念融入徽派文化，美食、美景、休闲、娱乐、文化一一囊括，让游客体验到舒适、别具风情的文化休闲享受。

Operation Measure 运营措施

The operation at the later stage is added with culture features. The reappearance of sea of clouds on Huangshan and the flash MTV on the wall make visitors feel like walking through the beautiful nature of the ancient town. There are also plenty shows for visitors. The street is just like a stage which lets visitors feel the happy atmosphere and enjoy the rich commercial activities of the street.

在后期的运营上，增添文化亮点，有黄山云海的再现、墙面音乐动画的呈现，漫步其中，仿佛置身于灵动仙境的古镇山水之间。同时为游客筹备丰富的娱乐节目，如古戏台的徽剧、傩舞、徽州婚嫁互动演出等。化街区为舞台，让所有到此的游客除了体验丰富的业态之余，还能感受到街区无比欢乐的氛围。

Gaochun Street, Nanjing
南京高淳老街

街区背景与定位
Street Background & Market Positioning

History 历史承袭

Built in Song Dynasty, its original name was Zhengyi Street. After the victory of the Revolution of 1911, its name was changed into Zhongshan Street to remember Sun Yat-sen. It was named Heping Street when Japanese troops invaded Chunxi Town. During the Cultural Revolution it was renamed Dongfanghong Street. It regained the name of Zhongshan Street in 1982.

Gaochun Street is a street of Ming and Qing dynasties and is preserved most completely in Jiangsu Province. It is famous for its antiquity and fancy. It is a national 4A tourist scenic spot and provincial level culture relic protection unit.

高淳老街始建于宋代，原名正仪街，辛亥革命胜利后，为纪念革命先驱孙中山先生，改名为“中山大街”。日军侵占淳溪古镇后，改称“和平街”。抗日战争胜利后，复名“中山大街”。“文革”期间，更名为“东方红大街”。1982年地名普查时，重新复名“中山大街”。

高淳老街是中国古街一颗灿烂的明珠，是江苏省保存最为完整的明清古街。古街以其古老、奇特而闻名，是国家AAAA级旅游景区、省级文物保护单位，被誉为“金陵第二夫子庙”。

Location 区位特征

Gaochun Street is located at the southwest of Chunxi Town, Gaochun County, Nanjing City. It is the most important street built adjacent to Guanxi River in Chunxi Town. Guanxi River is the main river way of Chunxi Town which is not only a significant communication between Taihu Lake and Changjiang River system, but also a distributing centre and economic artery of trades between Southern Jiangsu and Southern Anhui.

The geographical factors decided the important economic status of Gaochun Street and bred its unique culture.

高淳老街位于南京市高淳县淳溪古镇的西南部，老街是淳溪古镇最重要的街区，紧邻官溪河而建。官溪河是淳溪镇的主河道，既是沟通太湖和长江水系的重要水道，又是苏南与皖南进行大宗物资交易的集散地和经济命脉。淳溪古镇地处苏、皖两省交界之处，自古就是“吴头楚尾”，是连接苏南与皖南两大区域的水陆要道和重要的文化走廊。

吴头楚尾、湖河环绕的地理因素奠定了高淳老街重要的经济地位，孕育了其独特的地域文化。

Market Positioning 市场定位

Gaochun Street has a long history, thus there are cultural relics and historic buildings everywhere. Meanwhile it has a prosperous business with numerous time-honored shops. Therefore, the street is positioned as a historic culture street of traditional connotation and local features assembling shopping, tourism, culture, leisure and food.

高淳老街历史悠久，文物古迹、历史建筑随处可见；商业繁盛，街内店铺林立，老字号商铺遍布各处。老街的定位是将其打造集购物、旅游、文化、休闲、美食于一体的具有传统内涵和地域特色的历史文化街区，使其成为高淳人民引以为豪的生活家园，成为海内外游客感受文化、追溯历史、重温传统的“江南圣地”。

Street Planning 街区规划

In order to protect and restore the original appearance of Gaochun Street, till the end of 2008, the government of Gaochun County had invested over one hundred million RMB. According to the protection-oriented policy, the government strengthens the management of Gaochun Street and protects the existing layout, and adopts measures to gradually recover the street's best appearance of Ming and Qing dynasties. The measures are as follow:

From 1986 to 1992, the government totally invested more than one million RMB to preliminarily repair the façade of Zhongshan Street.

From 1993 to 1998, the goverment renovated all public facilities that did not meet with old buildings in the street, and invested 600 thousands RMB to repair the Wu's Ancestral Hall.

In 1999, the government carried out Temporary Provisions of Chunxi Street Management to encourage old shops to sell traditional and featured commodities and to forbid damage of original layout and appearance of old buildings.

In 2000, the goverment invested over three million RMB to restore Yang's Hall.

From 2001 to 2002, the goverment repaired Site of Gaochun Office of the New Fourth Army and Site of Church of Jesus.

In 2004, the goverment invested about nine million RMB to construct parking lots to forbid all motor vehicles from entering the street.

In 2008, Gaochun county party committee and government upgraded the management committee of the street to expand its limits of managing authority and strength.

In its future planning, the street will continue the protection-oriented policy to protect its historic reality, completed appearance and life continuity and finally to recover the best conditions of the old street of Ming and Qing dynasties.

为保护和全面恢复高淳老街的历史原貌，至2008年年底，高淳县政府累计投入资金1亿多元，坚持按照“保护为主、抢救第一、合理利用、加强管理”的方针和“修旧如旧”的原则，对高淳老街加强管理，保护街区现有格局，并且采取措施逐步恢复明清时期的最佳风貌。采取的主要措施如下。

1986年至1992年，累计投资100多万元，完成中山大街沿街立面的初步整治和恢复工作。

1993年至1998年，整修老街内与古建筑不匹配的市政设施，将老街上所有的电力、通信、供水等管线入地，并投资60多万元修复吴氏宗祠。

1999年，县政府制定实施了《关于淳溪老街管理的暂行规定》，鼓励老街商铺经营传统、特色商品，禁止破坏古建原有格局，要求店铺不得悬挂与古建筑不协调的现代招牌等，并专门成立了老街管委会。

2000年，投入资金300多万元，修复杨厅。

2001年至2002年，完成新四军驻高淳办事处旧址、耶稣教堂等5处古建的整修和环境保护。

2004年，投资近900万元，完成停车场的建设，从而为禁止一切机动车辆进入老街创造了条件。

2008年，高淳县委、县政府将老街管委会升格，使之独立开展工作，扩大其管理的权限与力度。

在未来的规划中，老街将秉承“保护第一、恢复为主、加强管理、合理利用、可持续发展”的基本原则，以保护老街的“历史的真实性、风貌的完整性、生活的延续性”为规划的核心，最终使老街恢复至明清时期的最佳状态。

Street Design Features 街区设计特色

Gaochun Street is a riverside street with a layout of cross lanes. It is composed of eleven streets and lanes including Old Street, Pawnshop Street, Chenjia Lane, Wangjia Lane and so on. It was 1,135 meters long in Ming and Qing dynasties, and now only 505 meters long are preserved. It is about 3.5 meters wide with grey stones vertically paved on both sides and rouge stones horizontally arranged in the middle.

There are many buildings of Ming and Qing dynasties style in this street. They are of distinct Anhui style. These old buildings along the river have a exquisite decoration like delicate sculptures and traditional calligraphy plaques. They are antique and elegant, therefore praised as "the epitome of oriental culture" by scholars of home and abroad and visitors.

高淳老街是一条纵横相交、完整分布的临河型街区。它由老街、当铺街、陈家巷、王家巷等11条街巷共同组成。在明清时期，老街全长达1 135米，现在保留下来的约505米，街面宽3.5米左右，街道两边用青灰石纵向铺设，中间用胭脂石横线排列，整齐美观，色调和谐。

高淳老街街区分布着成片的明清建筑，建筑一般为楼宇式双层砖木结构，挑檐、斗拱、垛墙、横衍、镂窗齐全，造型别致，古朴华丽，带有明显的徽派风格。这些古建筑傍水而列，粉墙青瓦、飞檐翘角，配上精美的砖木石雕和传统的书法牌匾，古朴典雅，被中外学者和游客誉为“东方文明之缩影”、“古建筑的艺术宝库”。其中规模大、特点鲜明的有吴家宗祠、杨厅、关王庙等。

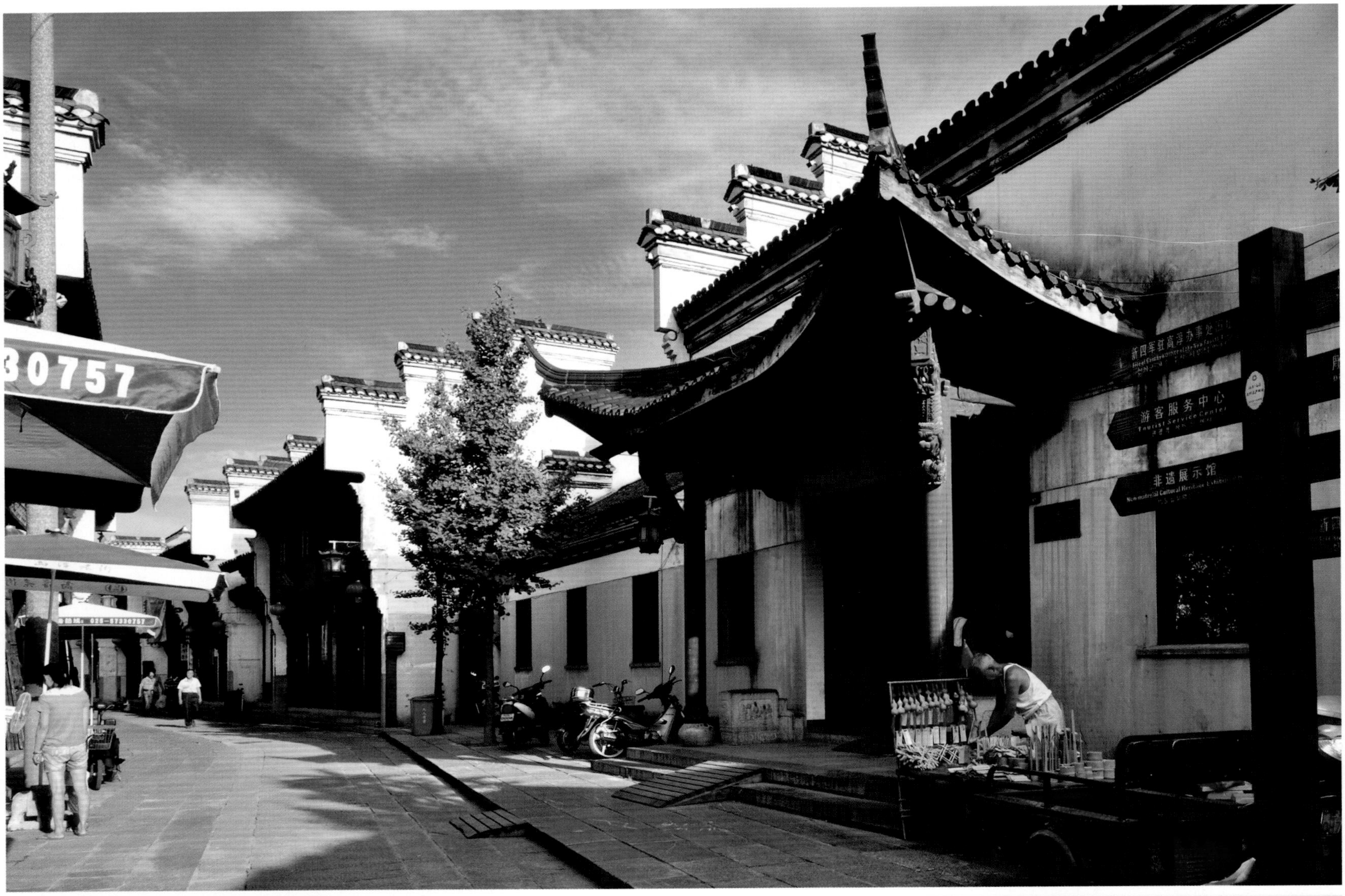
30757
游客服务中心
Tourist Service Center
非遗展示馆

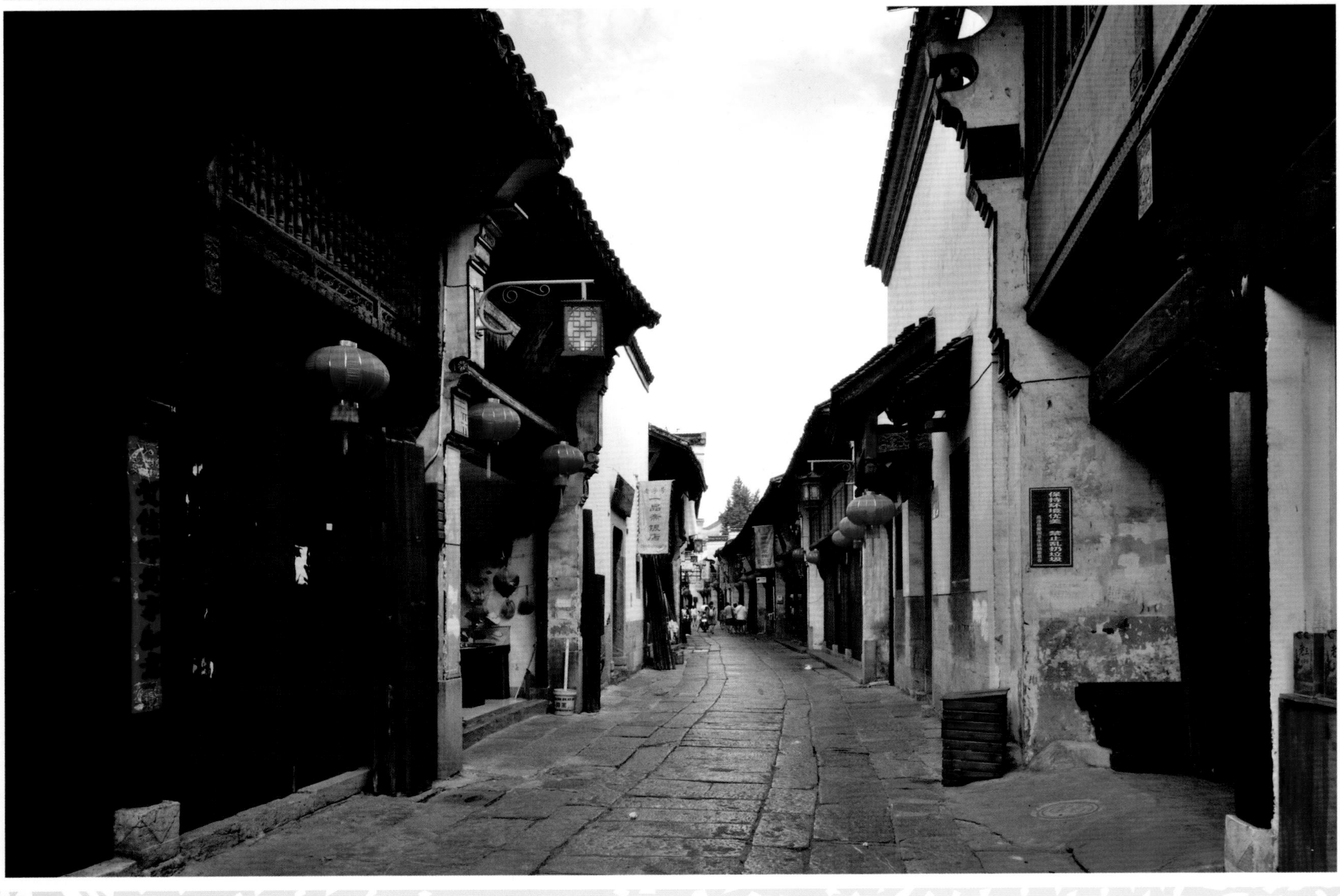

迎薰门美食一条街
老街桂花美酒十里飘香

迎薰门酒楼
耶稣教堂旧址

主入口广场
Main Entrance Plaza
耶稣教堂旧址
卫生间
TOILET
新四军一支队司令部旧址
雕刻展示馆
禁止车辆通行

Commercial Activities

商业业态

Major Commercial Activities 主要商业业态

Gaochun Street has a prosperous business since ancient times. Since Song Dynasty, it had been a collecting and distributing center of commodity trade between Jiangsu and Anhui. From Ming and Qing dynasties and the Republic of China, the business of the street was expanded and there were over 170 shops. During Guangxu Period of Qing Dynasty, there are eight famous drugstores.

The tradition of doing business in Gaochun Street has been passed down, thus the commercial atmosphere is still very strong in the street. At present, there are over 300 shops which are mainly traditional business, handicraft industry, restaurants and modern tourism. These shops not only offer local citizens daily necessities, but also provide visitors with local specialties. The featured commodities in Gaochun Street are famous all over the world. In addition, there are also crafts like pearl accessories, and delicious food.

高淳老街自古商业发达，自宋代起，即为粮食、布匹、茶叶、山货、药材等苏、皖两地大宗物资商品交流、转运的集散地。明清至民国时期，老街内有经营竹木材、茶叶、黄烟、药、盐、水果、炒货、羽毛贡扇、纸张等各色物品的店铺170余家。清光绪年间，老街上有名的药店就有8家，当时曾流传一段顺口溜："药材仁成堂，中街天兴祥，人参魏长庆，老店王元昌，胡家卖药连处方，夏家蚌壳疮药一扫光。"老街商业繁茂可见一斑。

高淳老街重商、经商的传统流传至今，老街的商业氛围依然十分浓厚。目前老街内有商家店铺300余家，主要以传统商业、手工业、餐饮业和现代旅游业为主。这些商铺既为本地人提供了各种日常生活用品，又为外地游客提供地方特色产品。老街之中的特色商品闻名于海内外，高淳羽扇明朝时期曾是皇宫贡品，世界钢琴大师查理德曾慕名前来购买高淳布鞋，除此之外，老街中还有珍珠饰品、玉泉炻器等充满地域特色的工艺品和香干豆腐、风味糕点、玉溪香鹅、固城螃蟹之类的美食。

Haorenjia Restaurant

Haorenjia Restaurant located at No.128 Gaochun Street is a local food restaurant featuring dishes of Gaochun flavor.

好人家菜馆

好人家菜馆位于高淳老街128号，是一家土色土香的土菜馆，主要经营具有高淳特色的地方土菜，特色菜有固城湖大闸蟹、椒盐小杂鱼、香干豆腐等。

Liuchaoju Hotel

Liuchaoju Hotel located at the west side of Gaochun Street is a hotel of over one hundred years old. It is of Anhui style with front and back two houses and a courtyard at the middle. The interior is decorated with rockeries, fountain and goldfish pond, and is comfortable and pleasing.

The melon feast of Liuchaoju is very famous there. There are dozens of dishes having a "melon" in their names.

六朝居大酒店

六朝居大酒店位于高淳老街西侧，是家百年老店。房屋粉墙黛瓦，精雕细刻，具有徽派建筑风格。房屋分前、后两进，中间有天井，寓意肥水不流外人田。屋内有假山、喷泉、金鱼池，舒适宜人。

六朝居的瓜字宴在当地十分有名，以“瓜”字命名的一桌菜有十几道，如子虾拌黄瓜、肉丝炒高瓜、火腿炖冬瓜……后来六朝居又增添了不少新菜，如固城围锅仙、慢城土鸡煲等。

Guanwang Temple 关王庙

Guanwang Temple is a temple worshipping Guan Yu. Guanwang Temple is an imitated antique building for worship ceremony and landscape appreciation.

关王庙是淳邑关羽之庙，始名关王庙，又称关帝庙、武庙、关岳庙。关王庙是一座既是祭祀又是景观性的仿古建筑，占地3 800多平方米，庙内建山门、照壁、戟门、祭殿（享殿）、启圣殿（正殿）、东西垛殿、钟鼓亭、“气肃千秋”“义贯云天坊”等坊。

Wu's Ancestral Hall 吴氏宗祠

Wu's Ancestral Hall built in 1781 is of brick and wood structure and Anhui style. It is divided into front, middle and back parts. The front part is a theatrical stage, the middle part is an ancestral hall and the back part is a sacrifice hall. In 1938, the New Fourth Army First Detachment led by Chen Yi set its command here. Now, the Wu's Ancestral Hall is a local custom museum.

吴氏宗祠始建于乾隆四十六年（1781年），为砖木结构，皖式风格，分前、中、后三进。第一进为戏楼，中进是享堂，后进为祭堂。1938年，由陈毅率领的新四军一支队的司令部就设在此处，在此期间，陈毅曾留下了《东征初抵高淳》的壮丽诗篇。现在吴氏宗祠已改为民俗馆。

Yang's Hall 杨厅

Built in the early Republic of China, Yang's Hall is vertically composed of three parts of 500 m². The front part is a two-floor house with three rooms in width. The middle and back parts are of Zoumalou style. There is a courtyard between each part for ventilation and daylight. There is a stone door sill before each hall meaning keeping away from misfortune and saving wealth. The decoration of the back part is different from Anhui style which is externally luxury; instead, it has a characteristic of Jiangsu style which is for self appreciation.

杨厅始建于民国初年，纵深为三进，占地面积500平方米。首进是一座面宽3间、上下2层的砖木结构楼房。中、后二进为“走马楼”造型。中进左右边间为主人及公子卧室，楼上则为小姐闺房。天井两侧设厢房，为来客和仆役居住。进与进之间设天井通风采光。每进堂间看柱内设石门坎，共3进4门，石门坎上安置6扇镂空屏风与门正对，意为对外邪不入内，对内财不外流。头进门面为排门式，门额之上设骑楼，檐口由曲缘外挑。后进雕花门罩朝内，上雕“德乃福基”，这区别于徽派华丽对外的特点，而具有苏派对内自观的特征。

Sculpture Gallery 雕刻展示馆

The Sculpture Gallery in Gaochun Street is divided into hall, wood carving area, stone carving exhibition area, tile carving area and Han portrait brick exhibition area. All the exhibits are collected from place to place in Gaochun County.

雕刻展示馆全馆共设门厅、木雕区、石刻展览区、砖雕区和汉画像砖展示区几大部分，展馆内所有展品都是从高淳县各地收集而来。

Old Street Art Club 老街艺术会馆

It is an important cultural place in Gaochun Street. It holds exhibitions of calligraphies, paintings and seal cuttings from time to time.

老街艺术会馆是高淳老街重要的文化场所，会馆内经常举办各种书画、印章篆刻等展览活动。

耶稣教堂旧址
Site of Church of Jesus
停车场
卫生间
TOILET
雕刻展示馆
Display Room of Sculpture
淳古菜
35

107
主入口广场
Main Entrance Plaza
北入口广场
North Entrance Plaza
新四军一支队司令部旧址
乾隆古井
Qianlong Ancient Well
新四军驻高淳办事处旧址
杨 厅
Yang's Hall
耶稣教堂旧址
Site of Church of Jesus
雕刻展示馆
Display Room of Sculpture

高
淳
老
街
耶稣教堂旧址
Church of Jesus
新四军驻高淳办事处旧址
仓巷
老字号

South Arrest Office Historic Culture Blocks, Nanjing

南京南捕厅历史文化街区

街区背景与定位 Street Background & Market Positioning

History 历史承袭

South Arrest Office Historic Culture Blocks is one of the existing building groups of Ming and Qing traditional style in Nanjing. According to the record, there were two government offices specializing in arrest in Nanjing when Qing Dynasty. The south one was engaged in land arrest, and the north one was responsible for arrest overwater. In the 1860s, the original building of South Arrest Office was ruined in Taiping Rebellion. The office was rebuilt in 1872, but a police office was set up to replace the arrest office in the late Qing Dynasty. Before the Anti-Japanese War the building of the office had disappeared, however the old street named after it was preserved.

南捕厅历史文化街区是目前南京老城内现存的明清传统风格建筑群之一。据史书记载，清代南京城内有南捕通判衙署（简称南捕厅）和北捕通判衙署（简称北捕厅）两个专门从事缉捕工作的衙署。东起府西街（今中山南路）、西接绫庄巷的南捕厅负责辖区内陆上的缉捕工作；位于府北土街口的（今中山东路、洪武路口的旧称）北捕厅则负责水上缉捕工作。19世纪60年代，原先南捕厅的建筑物毁于硝烟弥漫的太平天国运动中，1872年重建，清末实行新政时，又在此设立警察局取代了捕厅工作。到了抗日战争前夕，浮浮沉沉的南捕厅旧址房屋已荡然无存，但以此命名的老街巷却保存了下来。

Location 区位特征

South Arrest Office Historic Culture Blocksis locatd at the south of old Nanjing, connecting with city center and Confucian Temple. Its traffic is convenient with a completed road network. It has an advantageous location. It is the third phase project of South Arrest Office historic protection zone.

南捕厅历史文化街区位于南京老城南部，连接市中心与夫子庙两大商业圈，与新街口相距仅1 000米，周边道路网络完善，出入便捷，紧邻地铁1号线三山街站。区位优势显著，是南捕厅历史风貌保护区建设的第三期项目。一期工程修缮了甘熙故居，二期工程建成了目前的“熙南里”街区。

Market Positioning 市场定位

By constructing and remolding the blocks to create a cultural tourism commercial street of traditional style which integrates culture, history and street.

通过对街区的建设与改造，形成集文化、历史、街区为一体的传统风貌文化旅游商业街区。

To inherit Jinling culture and show its life style. With a staring point of remolding environment and inheriting history, to deeply explore the historic culture connotation of the city, thus to enhance the city's historic culture connotation and manifest its historic culture features.

街区的开发理念是传承金陵文化，演绎风尚生活，以改造环境、传承历史为出发点，深入挖掘城市历史文化内涵，提升城市历史文化底蕴，彰显城市历史文化特色。

Street Planning 街区规划

According to its general principal of "respecting and preserving tradition, applying modern material and technology", four stages are planned.

The First Stage of Planning (2002 to 2003)

Function positioning: strengthen the demand of traditional protection and modern development and make it a historic culture protection zone.

Development principle: keep the basic fund balance of protection, removal and construction of the planning.
Preserved buildings: 9 first type buildings, and 33 second type buildings.

以"尊重传统城市肌理，保留传统历史文脉，沿袭传统建筑空间，应用现代材料技术，适应现代生活模式，提升现代城市活力"为总体原则，规划经历了四轮的逐步过渡。

第一轮规划（2002—2003年）

功能定位：坚持传统风貌保护与现代化发展的要求，以传统街巷肌理为核心，成为融特色居住、旅游、文化、休憩等功能于一体的历史文化保护区。

开发原则："就地平衡"，依据当时的拆迁补偿费用测算，方案中的保护、拆迁、建设等资金基本平衡。

保留建筑：保留一类建筑9处，二类建筑33处，共计42处。

The Second Stage of Planning (2006)

Background: after the implementation of new remove policy, the economic balance can't be kept any more. Because of lack of clear protecting management, many buildings will be damaged more severely after years. What's worse, residents' private buildings will further destroy the building style.

Scheme: according to the economic condition, three schemes are planned to choose. The first one is to improve plot ratio by high-rise buildings; the second is to take pure residential development as the principal thing; and the third is to realize the style protection as much as possible by limited high-rise buildings.

Development principal: locally principal.
Preserved buildings: 9 first type buildings, and 7 second type buildings.

The Third Stage of Planning (2007 to 2008)

Background: the knowledge of historic culture resources of different fields from society has changed.
Preserved buildings: 17 first type buildings, and 12 second type buildings.

The Fourth Stage of Planning (2009 to now)

Scheme: according to the principal of "protecting as much as possible" put forward by professors, the former scheme is improved and adjusted. The protection of traditional buildings is classified into four kinds: entirely protection, partial building protection, part decorating protection and form and texture protection.

Preserved buildings: 68 planned protection buildings of higher value, 43 partial protection buildings.
A general layout of "one core, two axes, one area and one district" is planned. "One core" refers to Ganxi Mansion "Two axes" means a history exhibiting axis and a folk custom exhibiting axis "One area" means a business service area; and one district means a complex district of business and office.

第二轮规划（2006年）

制定背景：新的拆迁政策实施后，街区拆迁成本已经超过每平方米7 000元，无法形成街区内经济建设的“就地平衡”。因为没有明确的保护管理要求，原定的保留建筑多为木结构，经过几年时间后，许多建筑自身老化毁坏更加严重，居民的私搭乱建行为进一步加剧了建筑风貌的毁坏。

方案情况：根据经济条件，分别制定了三个方案：一是以高层建筑来提供容积率；二是以纯住宅开发为主；三是以有限的高层建筑来实现最大限度的风貌保护，强调整体格局保护，提高居住形态和服务业态的档次。

开发原则：就地平衡。

保留建筑：保留一类建筑9处，二类建筑7处，共计16处。

第三轮规划（2007—2008年）

背景：社会各界对历史文化资源保护的认识较以前发生了很大变化

保留建筑：保留一类建筑17处，二类建筑12处，共计29处。

第四轮规划（2009年至今）

方案情况：根据专家提出的“能保则保”原则以及多元化保护思路，对方案进行了完善和调整。在现场重新踏勘调查的基础上，扩大历史文化保护的内涵，将传统保留建筑的保护方法分为四类：整体保护、部分建筑保护、局部装修保护、形式与肌理保护。此外，根据今后的具体规划和地下开放的要求，对其中部分建筑进行移建或复建。

保留建筑：68处有较高价值的建筑进行拟保护，43处进行局部保护建筑，共计111处。

规划整体形成“一核、二轴、一带、一区”的布局。“一核”为全国重点文物保护单位甘熙宅第，从片区工坊制作技艺的创造性特征出发，继承和发扬南京的创意文化；形成以历史展示轴和民俗展示轴的“二轴”，来展示南捕厅街区所承载过的历史功能；“一带”是商业服务带；“一区”为商办混合区，在功能上与甘熙宅第相补充。

Street Design Features 街区设计特色

The design seriously considers two elements — historical context and developing tourism resource of former Ganxis residence. The blocks are divided into three parts, south part, north part and west part. The major function of south part is boutique shopping, restaurant, club and hotels. The north part is a building group along Pingzhang Lane. And the west part is four preserved buildings and urban greenbelt.

The compound mode of building spaces is mainly traditional courtyard with slope roofs and main colors of black, white and grey. What's more, great attention is paied to create a pedestrian space with a clear tourist route.

设计着重考虑延续历史文脉和开发甘熙故居旅游资源两大因素。街区总体上被分为3个区，分别为南区、北区和西区。南区的主要功能定位为精品购物、餐饮、企业会所和精品酒店。该区设有地下车库，提供235个停车位；北区为沿平章巷建筑群，其突出特征为以“一大数小”的内向庭院组织套内院落，由2～11户相连形成若干个建筑群落，各群落之间传承传统的肌理街巷分割，地下车库可提供61个停车位；西区为4栋保留建筑和城市开发绿地。

建筑空间组合方式以传统院落为主，坡屋顶，色彩上以黑、白、灰为主色调。并着重塑造南捕厅、大坂巷等街巷的步行空间，旅游动线清晰明了。

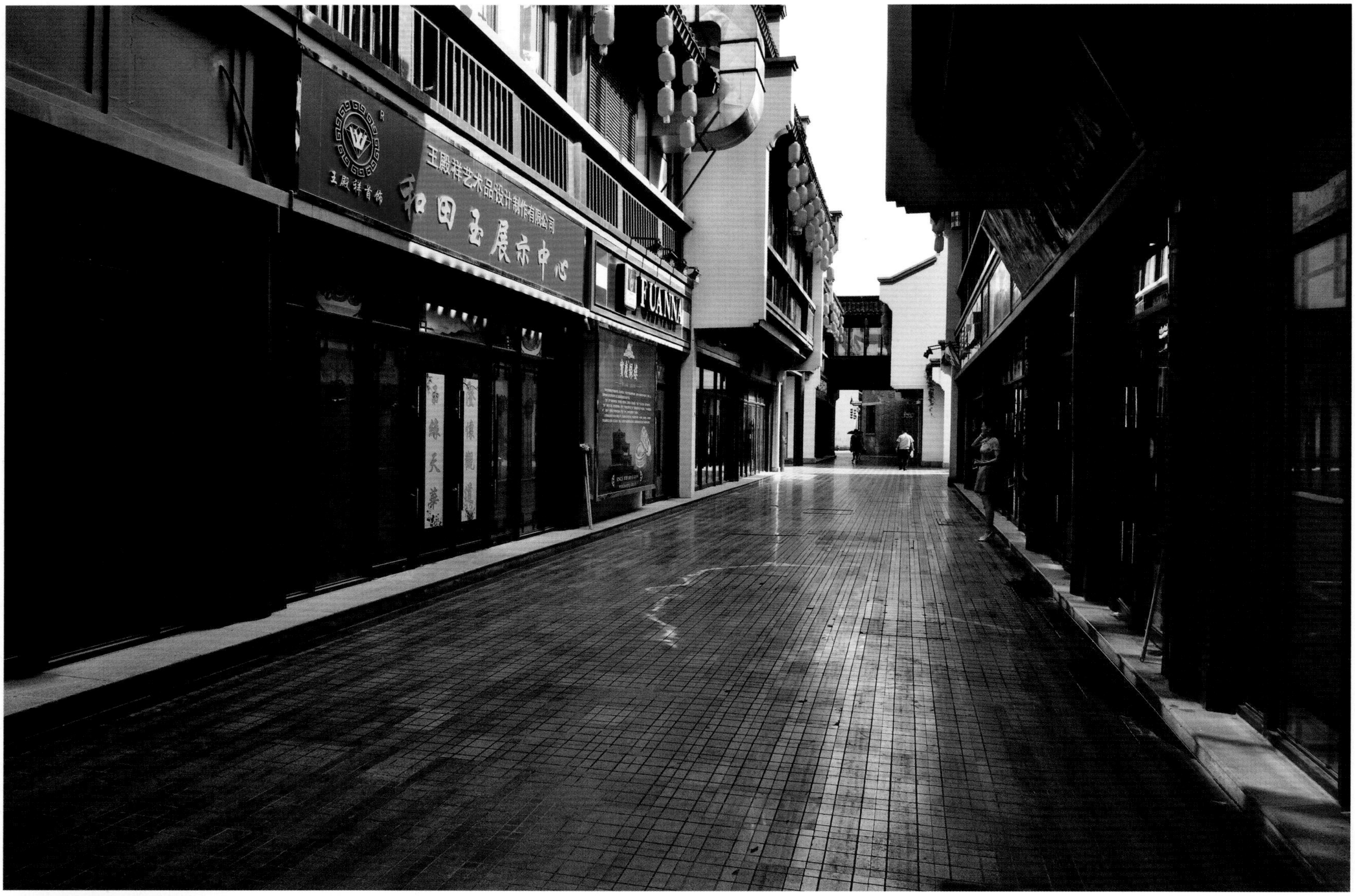
王殿祥首饰
王殿祥艺术品设计制作有限公司
和田玉展示中心
FUANNA

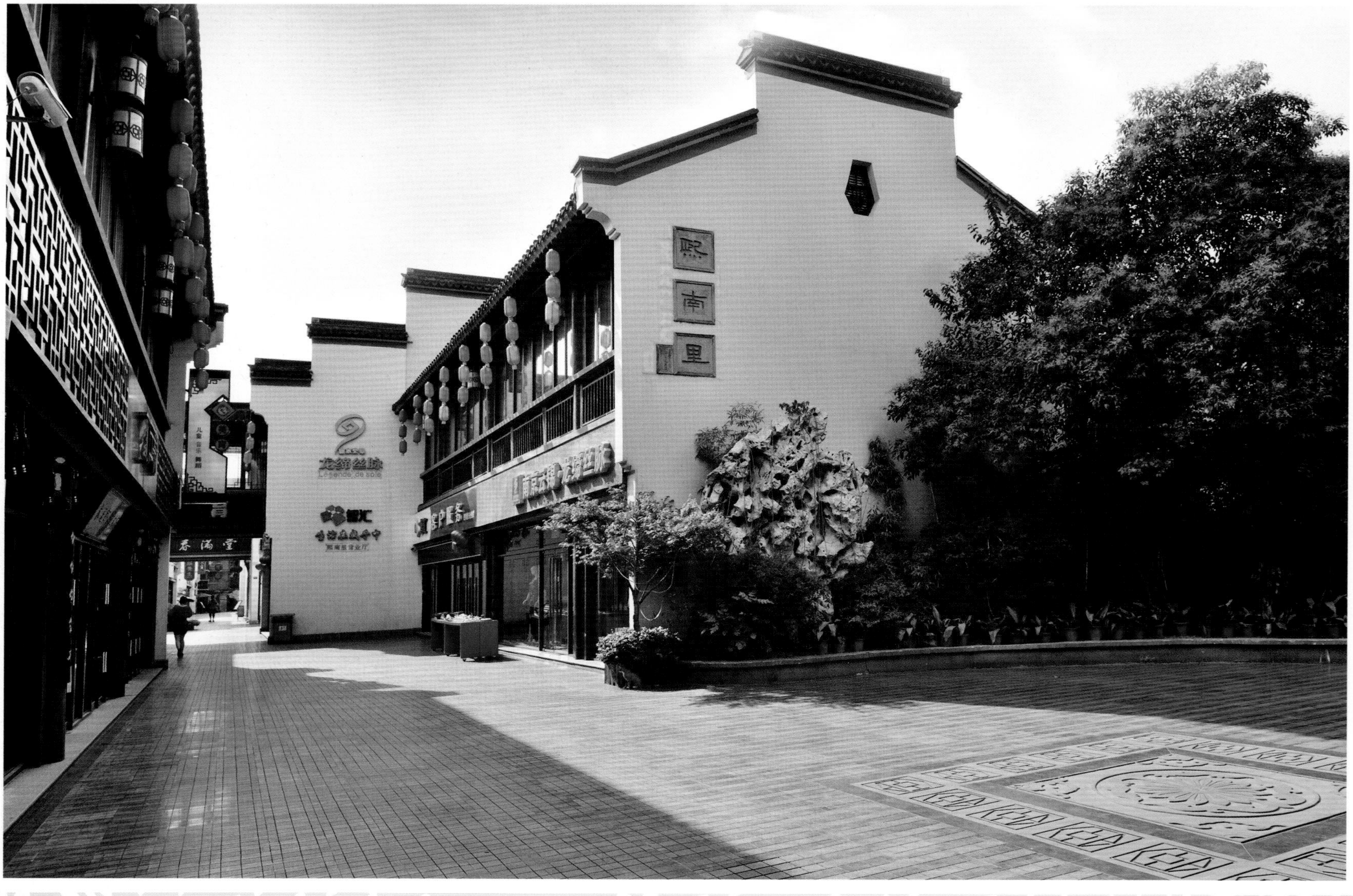
南
里
龙錦丝緣
香满堂

棋牌包间 私房菜馆
茗茶棋牌简餐
全国重点文物保护单位
甘熙宅第
中华人民共和国国务院
二〇〇六年五月公布

江宴楼
江宴楼
江宴楼
春满堂

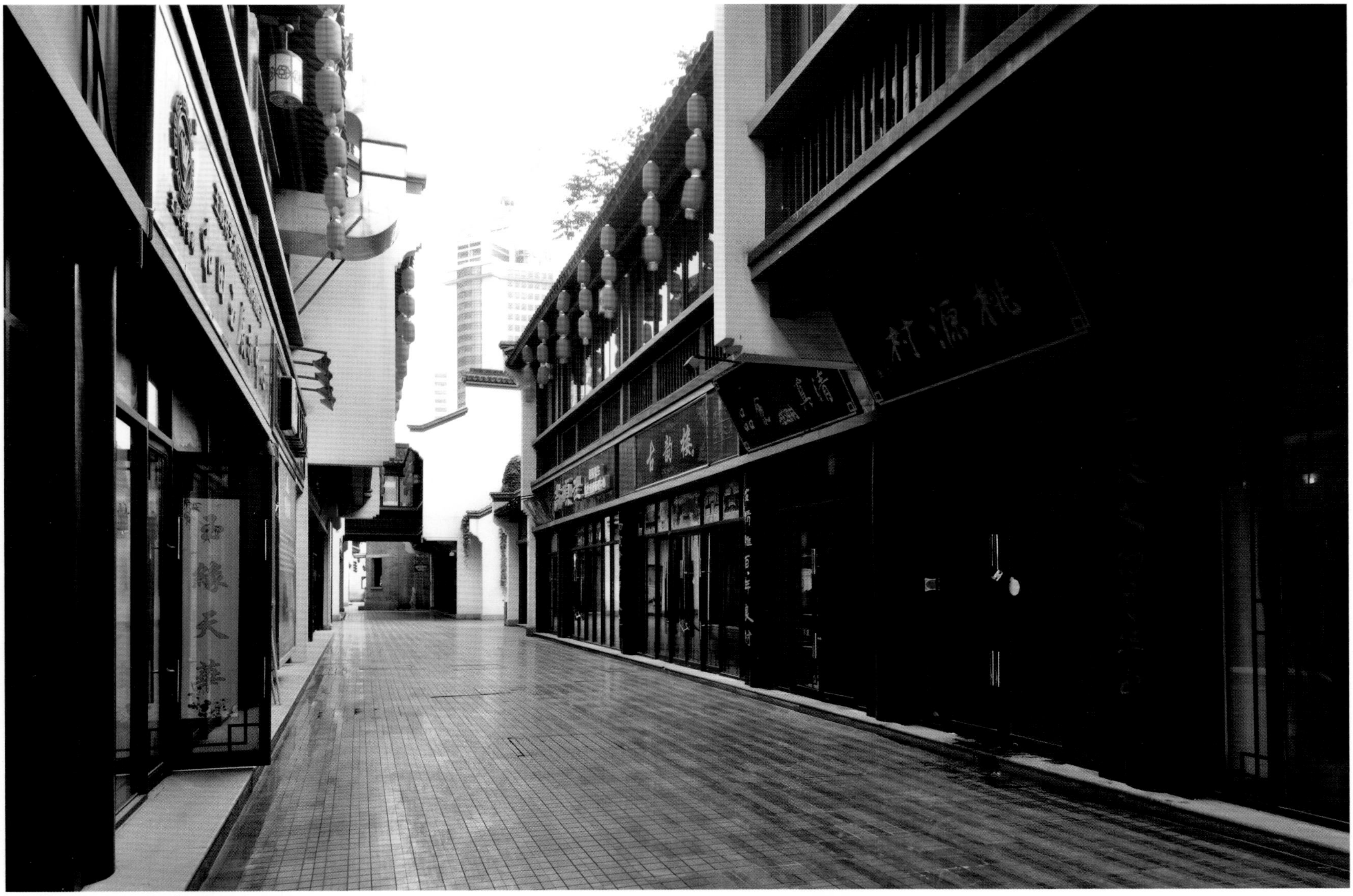
桃源村

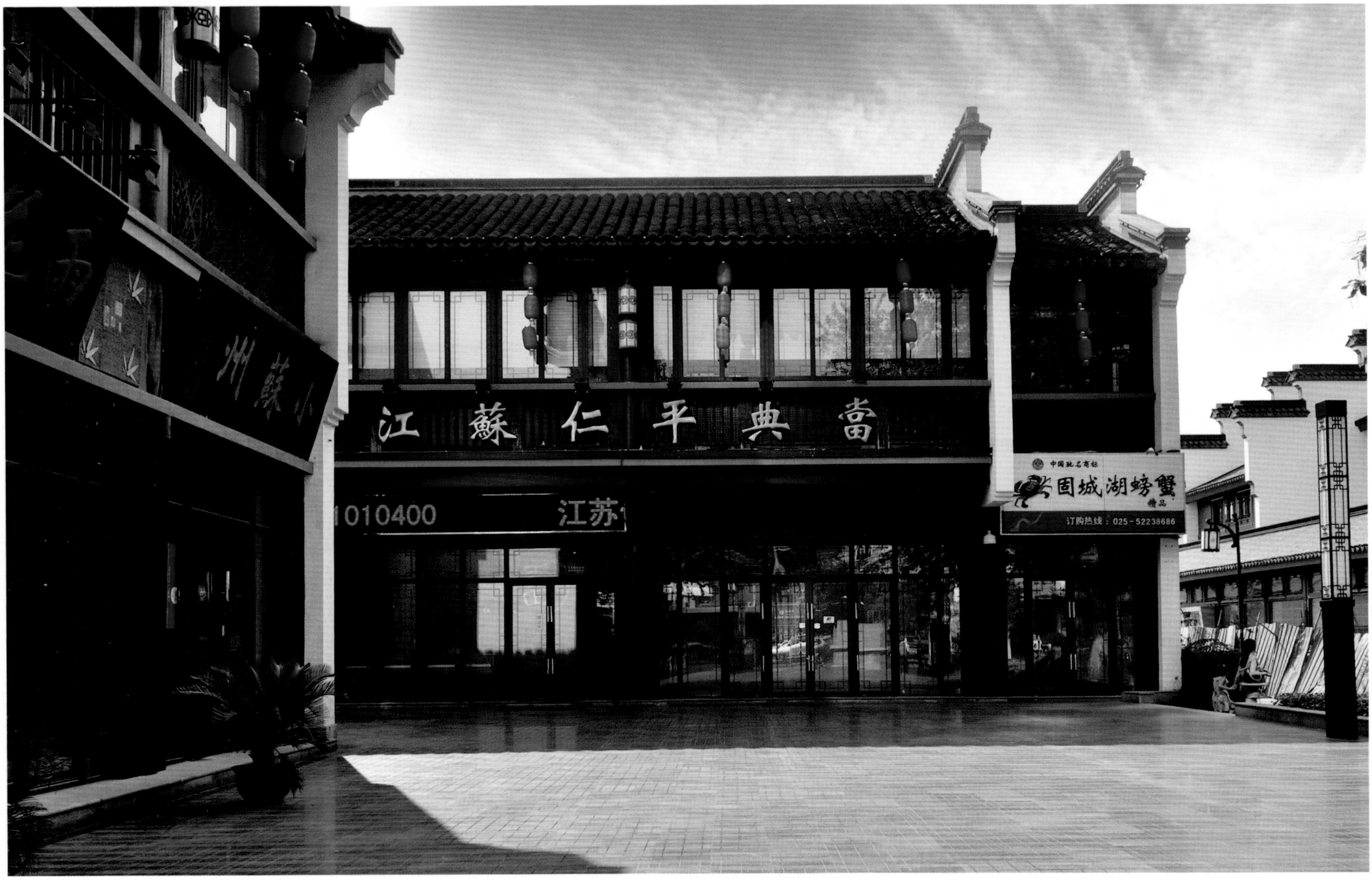
小蘇州
江蘇仁平典當
1010400
江苏
中国驰名商标
固城湖螃蟹
精品
订购热线：025-52238686

FINNEGANS WAKE
EST.2005

棋牌包间 私房菜馆
订座电话
雨花茶坊

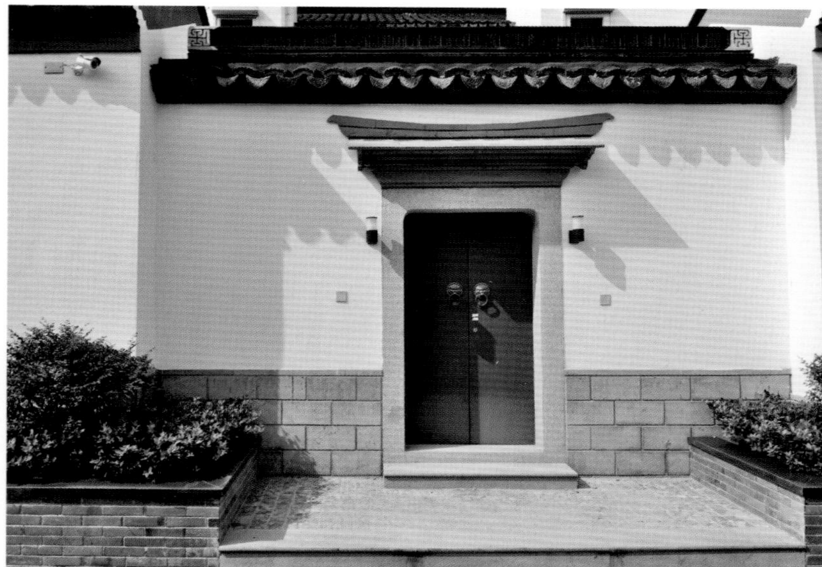

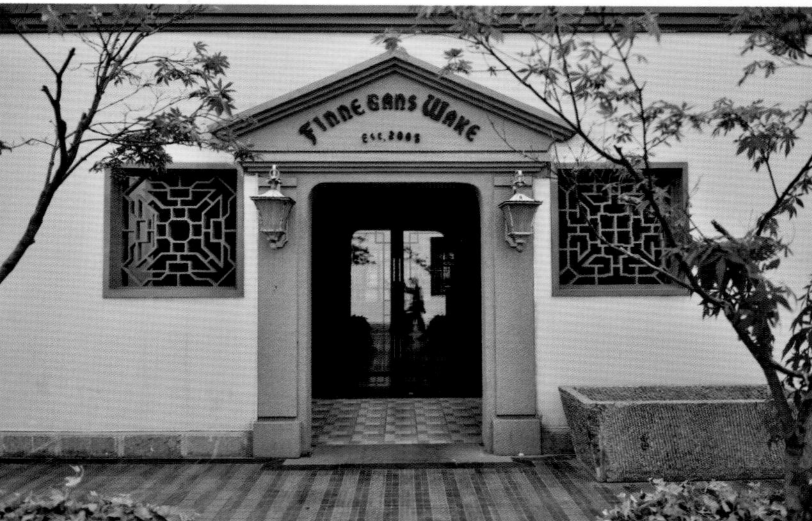

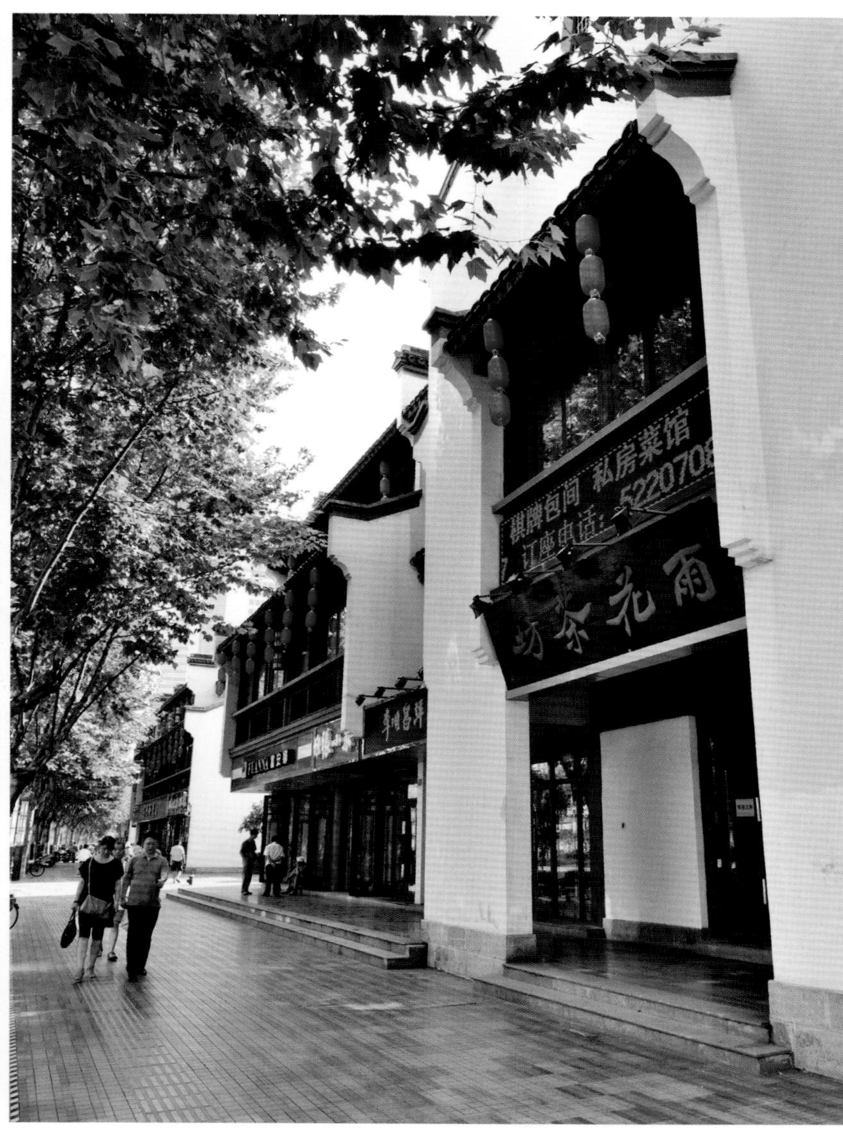

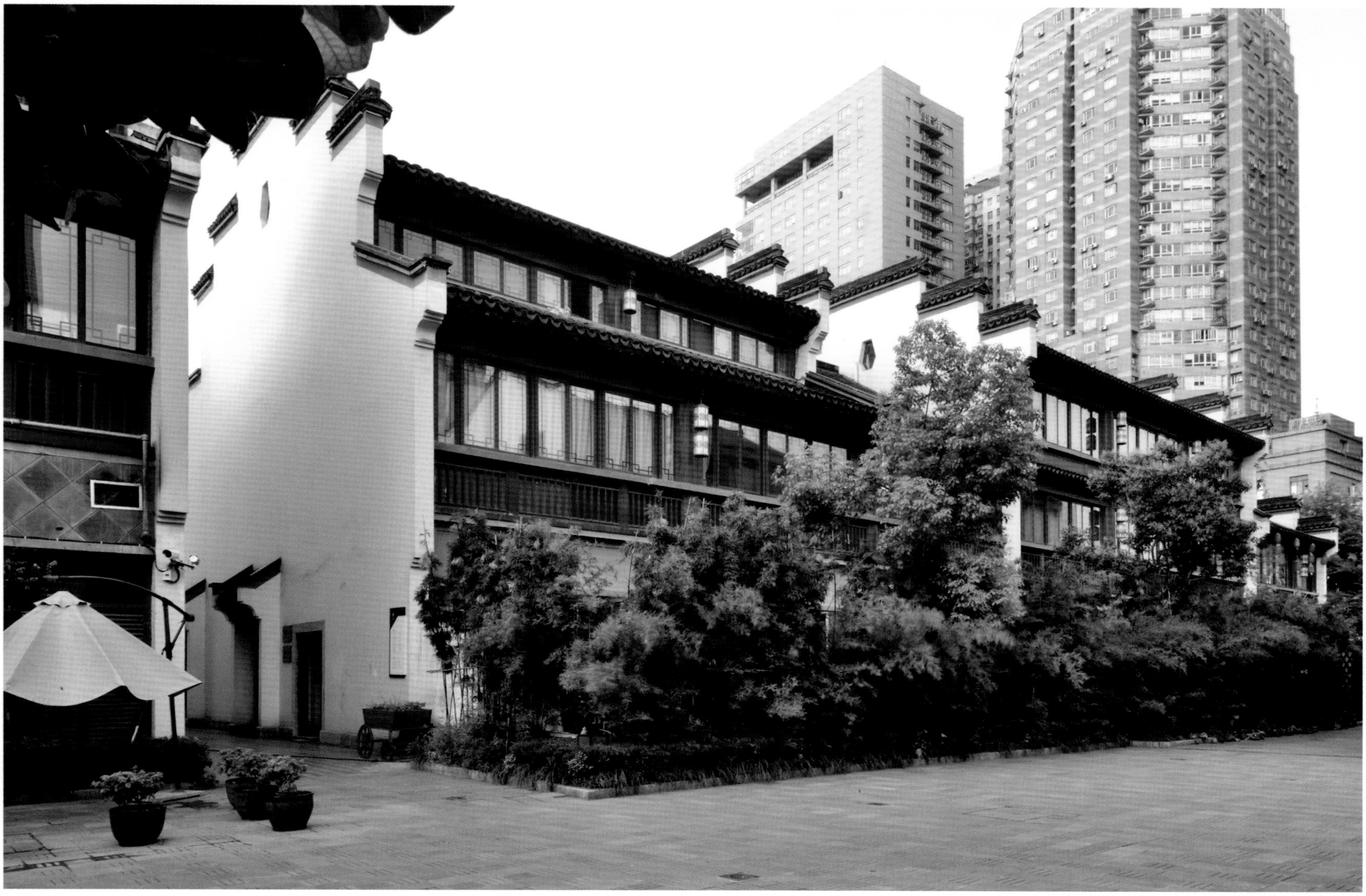

Major Commercial Activities 主要商业业态

Commercial activities here are mainly commercial supports taking Ganxi Mansion as a core with "Jiangnan Qi Shi Er Fang" as general positioning. Local cultural features and intangible relics are used to create business features of original ecology, multi-layer and living. Historical and cultural relics are protected and intangible culture remains are fully explored to form a unique business industry cluster district by molding building spaces.

主要商业业态是以甘熙宅第为核心的商业配套，占地面积123 000平方米，以“江南72坊”为总体定位，包含公益性博物馆坊、可经营的老字号坊、代表江南文化的记忆坊、传承物质与非物质文明的文化坊等。围绕“琴、棋、书、画”文化特色，结合地块内蕴含的丰富的非物质文化遗产，打造原生态、多层次、生活化的商业特色。在保护好历史文化遗产的同时，充分挖掘非物质文化遗产，通过建筑空间的塑造形成独具特色的商业产业集聚区。

Food & Beverage 餐饮类

Ayingbao

It is an AAA level green restaurant in China. It uses its own special seasoning and cooking method to maximally exert the fragrance and delicate flavor of the raw materials, therefore leading customers to endless aftertastes.

阿英煲

阿英煲是中国AAA级绿色餐馆，嘉兴市著名商标、全国连锁餐饮企业，把农家放养的土鸡、老鸭以及手打豆制品、腌品和各种绿色土特产配送到各连锁店。用独有的秘制调料和烹饪方法，不过油、不煎炸，通过焖、炖、焐等传统方法，最大程度发挥原料本身的香味和鲜味，使每道菜都保持原汁原味，且醇正味美，令人回味无穷。

Jiangyan Lou

Jiangyan Lou has a modern decoration concept and function, and at the same time inherits the style and charm of Ming and Qing dynastyies.

江宴楼

江宴楼是江宴渔村酒店有限公司倾心打造的商务宴请会所，既有现代的装修理念和功能，又传承了明清时期的格调和古韵，客人亦可享受顶级鲍鱼、鱼翅，凸显消费者的高贵与尊荣。

Lvliuju

Lvliuju Restaurant was founded in 1912 and its name was from its location among willows at the side of Qinhuai River. It was famous for its authentic vegetable dishes. After generations of chefs' elaboratel innovation, its dishes are not only of good looking and taste, but also further highlight "freshness, tenderness, hotness, crisp and fragrance" five characteristics.

绿柳居

绿柳居菜馆创建于1912年，因坐落于秦淮河畔桃叶渡的绿柳中，故名“绿柳居”。以经营正宗素菜而闻名遐迩，店堂墙上一幅“八仙过海”壁画，栩栩如生，仿佛在告诉食客，这八仙皆因吃素而得道。绿柳居的素菜上承六朝余绪，下应时令风尚，经过几代名厨的悉心创新，不仅菜肴造型逼真、口味独特，还进一步突出了“鲜、嫩、烫、脆、香”五大特色。

Leisure 休闲类

Waku

Waku is elaborately designed according to Chinese traditional buildings by Yu Ping, a famous designer in China. It is a tea art house of tile theme. It provides customers with over 60 kinds of famous tea in the country and simple Chinese cuisines. It also offers other beverages like coffee, milk shake and so on.

瓦库

瓦库是由国家著名设计师余平根据中国传统建筑精心设计、以瓦为主题的茶艺馆。瓦库以茶为经营核心，提供全国60余种名茶，另配中式简餐料理，还有咖啡、冰沙、奶昔等系列饮品。

Nianyi Club

Nianyi Club is a Jiangnan style building absorbing essential concept of Ming and Qing traditional courtyard. On the basis of protecting the historic building, modern technological elements was integrated to make the club grand and magnificent among other buildings.

廿一会所

廿一会所建筑为江南墅院，白墙黛瓦，吸收了明清传统庭院的精华理念，经过设计师巧妙改造，在保护历史古建筑的原址之上，融合现代科技元素，使“南京廿一”在众多建筑中别显大气恢宏。该会所汇集高档中餐厅、私家戏台、雪茄酒廊、多功能宴会中心于一体。

Zhang Xiaoquan

Zhang Xiaoquan scissors main store in Nanjing was founded by Zhou Mingzheng, an apprentice in "He Ji" Zhang Xiaoquan scissors store in Hangzhou. In 1933, Zhou Mingzheng founded his own store named "Zheng Ji" Zhang Xiaoquan scissors shop in Nanjing.

张小泉

南京张小泉刀剪总店创始人周明正（浙江省杭州萧山人）是杭州赵善康所开的"鹤记"张小泉刀剪店的学徒。周明正于1926年开始到南京发展，1933年在南京正式设店取名"正记"张小泉刀剪店。

Muyu Xuan

It sells classic furniture and crafts, jade articles and so on.

木玉轩

木玉轩经营黄花梨、紫檀、红酸枝、鸡翅木等古典家具、工艺品以及沉香、玉器、字画等。

Health Care 养生保健类

Chunmantang

It devotes to health care and advocates an initiative healthy concept of "prevention before disease onset". It inherits and promotes traditional "Chinese medical health maintenance" culture, and applies traditional Chinese medical essence to customize characteristic health care productions for consumers.

春满堂

春满堂致力于健康事业，倡导“未病先防、既病防变”的主动健康理念。店内经营诸多优质天然滋补、调养产品，如冬虫夏草、人参、鹿茸、西洋参、燕窝、锁阳、天麻、三七、当归、黄芪、枸杞等。传承和发扬传统的“中医养生”文化，运用传统的中医养生精髓为消费者量身定制个性化的养生产品。

Shangyisheng

Shangyisheng Health Management Center is a health service organization of prevention, health protection and treatment. It provides members with targeted guidance according to their life habits and environment.

尚颐泩

尚颐泩健康管理中心是融预防、保健、治疗于一体的健康服务机构。根据会员的生活习惯、生活环境进行有针对性的指导，使其得到个性化的医疗保健服务，足不出户就能解决日常健康咨询、健康方案、健康评定、养生保健等问题。

Nature Therapy & TCM

Nature Therapy & TCM originated in Beijing and absorbed "comfortable nursing" mode from Taiwan and Singapore. It provides customers with an healthy life style easy to persist.

望族

望族源于北京，吸收中国台湾、新加坡中医的“舒适调理”模式，汇集理疗、营养、减压调理和中医脾胃等领域的专家，回归中医“治未病”和自然疗法的根本，提供一种“易于持续的健康生活方式”。

Home Wares 家居类

Guoxiangguan

With natural fragrant materials as carrier, Guoxiangguan integrates natural science with humane studies to develop and spread fragrant product of healthy and fashionable elements.

国香馆

国香馆家居以天然芳香原料作为载体，融自然科学和人文科学为一体，致力于研制、开发、传播具有丰富健康时尚元素的香产品，创造和延伸精英人群的高端享受，并将其整合成为中华养生文化精粹与现代休闲生活结合的香道艺术，传承和发扬中国传统香文化。

Ganxi Mansion 甘熙宅第

Ganxi Mansion was built in Jiaqing Period of Qing Dynasty, and it, together with Tomb of Emperor Zhu Yuanzhang and Ming Dynasty City Wall, is called three top sights of Ming and Qing dynasties in Nanjing. The layout of this old building strictly follows the patriarchal idea and family system, therefore, the building has a grand scale and all of its location of rooms, decorations, sizes and shapes follow a unified grade rule.

甘熙宅第又称“甘熙故居”“甘家大院”，始建于清嘉庆年间，俗称“九十九间半”，与明孝陵、明城墙并称为南京明清三大景观，具有极高的历史、科学和旅游价值，是南京现有面积最大、保存最完整的私人民宅。整个建筑反映了金陵大家士绅阶层的文化品位和伦理观念。建筑的布局严格按照封建社会的宗法观念及家族制度而布置，讲究子孙满堂、数代同堂，致使宅第的规模庞大、等级森严，各类用房的位置、装修、面积、造型都具有统一的等级规定。

引导指示系统
Guidance & Sign System

CINNALANE
熙南里
江宴楼
Jiang Yan Lou
国香馆
Guo Xiang Guan
瓦库
Wa Ku
美育儿童
音乐舞蹈
春满堂
Chun Man Tang
望族养生
Nature Therapy&TCM
芬尼根酒吧
Finnegans Bar
木玉轩
Mu Yu Xuan
永为好
Yong Wei Hao
听秋阁
Ting Qiu Ge
江南七十二坊
天下玉家 SKYWORD
金熙楼
Jin Xi Lou

尚
頤
CINNALANE
熙南里

江南七十二坊
天下玉家SKYWORD
金熙楼
CINNALANE
熙南里

Guanqian Street, Suzhou
苏州观前街

街区背景与定位
Street Background & Market Positioning

History 历史承袭

Guanqian Street is an ancient street of more than one thousand years history. It was named Guanqian Street, for it was located in front of Xuanmiao Temple. Its name had changed according to the name of Xuanmiao Temple. It was renamed Zhongshan Road in 1937, Dongfanghong Street during the "Cultural Revolution" and Guanqian Street again in 1980. It was set for a pedestrian street in 1982.

Guanqian Street is one of the most prosperous commercial streets in Suzhou. It has profound historic culture and is an epitome of architecture and culture of Suzhou.

观前街是一条有千余年历史的古老街道，因其在千年名观——玄妙观前而得名。观前街的名字随玄妙观名字的变化而变化，曾被称过“天庆观前”“玄妙观前”，也曾因为玄妙观内桃花繁盛，灿若云锦，而被称为“碎锦街”。1937年时曾叫“中山路”，“文革”期间又叫“东方红大街”，1980年后恢复旧称“观前街”，1982年正式定为步行街，是国内最早的一批步行街之一。

观前街是苏州最繁华的商业街道之一，它有着厚重的历史文化积淀和特色，是苏州建筑和文化的缩影，直到现在观前街依旧是苏州商业繁华的区域。

Location 区位特征

Guanqian Street is located at the center part of Suzhou. It extends from Cufang Bridge Lindun Road at the east to Renmin Road at the west. The Xuanmiao Temple is sitting at the middle of the street. Being 780 meters long and 720 meters wide, the street has an advantageous geographical condition which makes it become one of the most prosperous commercial streets in Suzhou.

观前街街区地处苏州中心部位，东起醋坊桥临顿路，西接人民路，北靠旧学前、因果巷，南抵干将路，坐北朝南的玄妙观坐落于街道中段。观前街全长780米，街区南北宽720米，占地面积56.72公顷。优越的地理环境使观前街成为苏州最繁华的商业街道之一，成为苏州服务、文化中心。

Market Positioning 市场定位

The market positioning of Guanqian Street is to highlight the feature of "boutique well-known shop street" and to make it a top international modern commercial pedestrian street gathering shopping, leisure, restaurant, entertainment and tourism.

观前街的市场定位为：突出“精品名店街”的特征，将其建设成为以精品店、特色店、专业店为主，国内一流、具有国际先进水平的集购物、休闲、餐饮、娱乐、旅游为一体的现代化商业步行街。

规划设计特色

Planning & Design Features

Street Planning 街区规划

On the basis of analyzing demand of planning design, the planning aim of Guanqian Street is determined. The planning aim of Guanqian Street is to show and explore local features instead of blindly looking for innovation and gorgeousness, neither deliberately imitating antique. Guanqian Street should boldly show its style and features to guide Suzhou culture and present Suzhou features.

Therefore, the planning structure of Guanqian Street mainly emphasizes three functions: historic culture, business and sightseeing. The original planning layout of Guanqian Street is reorganized. Residences, administrations and restaurants are moved out to make the street specialize in commercial use. Tourism culture and entertainment facilities around the street are organized to add its attraction. Lands for business, cultural, greening and leisure use are expanded to highlight the theme of the street and inherit the former excellent features of Guanqian Street.

在对观前街进行规划设计需求分析的基础上，确定了观前街的规划目标是要展示和挖掘当地特色和苏州特色，不能一味求新、求大、求气派，也无须刻意地做旧、做古，在对旧的把握上要参照街道在改造以前最繁华并最能体现其完整风貌的时期，观前街应该大胆地展示其面貌，引导苏州文化，展示苏州特色。

因而，观前街区的规划结构主要突出了三个方面的功能：一是历史文化性，二是商业性，三是可游性。根据这个定位，对观前街区原有的规划布局进行了重新组织，将观前街内住房、行政、餐馆等场馆迁出，专做商业用途，在其周围组织旅游文化、娱乐等配套功能区域，增加街区的吸引力，使商业氛围增加。减少原来的居住用地、工业用地、行政办公用地，扩大商业贸易、文化古迹、道路广场、植树绿化及停车休闲用地，使整个街区主题突出，更具条理，既继承观前街过去的优秀特点，又通过更新连接了现代城市街道的便利性与可识性。

Street Design Features 街区设计特色

Guanqian Street is not just a straight street, but slightly diagonal and curve, so the street space continuously changes with memorial archways or raised platforms around wells as guidance. Besides, the original street space is also reasonably widened or narrowed. Some intersections and entrance spaces are widened or given a wider feeling by traditional archways, and the street space is narrowed by eave galleries or arcade buildings, therefore to strengthen the direction sense of the space.

The buildings which are of various styles at the street are antique, modern or of European style. The middle part of the street centralizing Xuanmiao Temple has a strong traditional building style; the east part presents a local style; the west part featuring modern buildings shows a age breath. Among these various style buildings, Chinese style buildings are the major whose heights and sizes are harmonious with surroundings. Their colors are mostly black matching white which could continue the Suzhou building's features in the modern environment.

在观前街的街道设计上，观前街并不是一条笔直的大道，而是斜线形、折线形或曲线形的大道，围合街巷的建筑也都有略微的凹凸进退的变化或角度的微小偏转。这使街巷空间不断地产生变化，再配合一些牌坊或井台之类的标志物以起到方向的引导性。另外，在街道的改造中也注意对原街道空间进行有节奏的收放处理：对邵磨针巷、大成坊、宫巷、临顿路等道路的交叉口和入口空间做了放大处理，或以传统牌坊的形式加强标识性和引导性给人以开敞感，街道行走空间则通过增设檐廊、骑楼来缩小空间，一收一放，产生了变化，增强空间方向感。

街区建筑或仿古，或现代，或欧化，建筑风格多样、自由。街道中段以玄妙观为核心，形成浓郁的传统建筑风格；东段以苏州地方风格为主，体现一定的地方风格；西段以现代建筑为主，体现时代气息。在多样化的风格中以中式建筑为主，注意了高度体量上与周围建筑的协调，建筑高度以3层为主。在色调上多选用黑白色调搭配，特别高大者以灰色调为主在形式上错落变化，将苏州粉墙黛瓦、轻灵小巧的感觉在现代环境中延续下去。

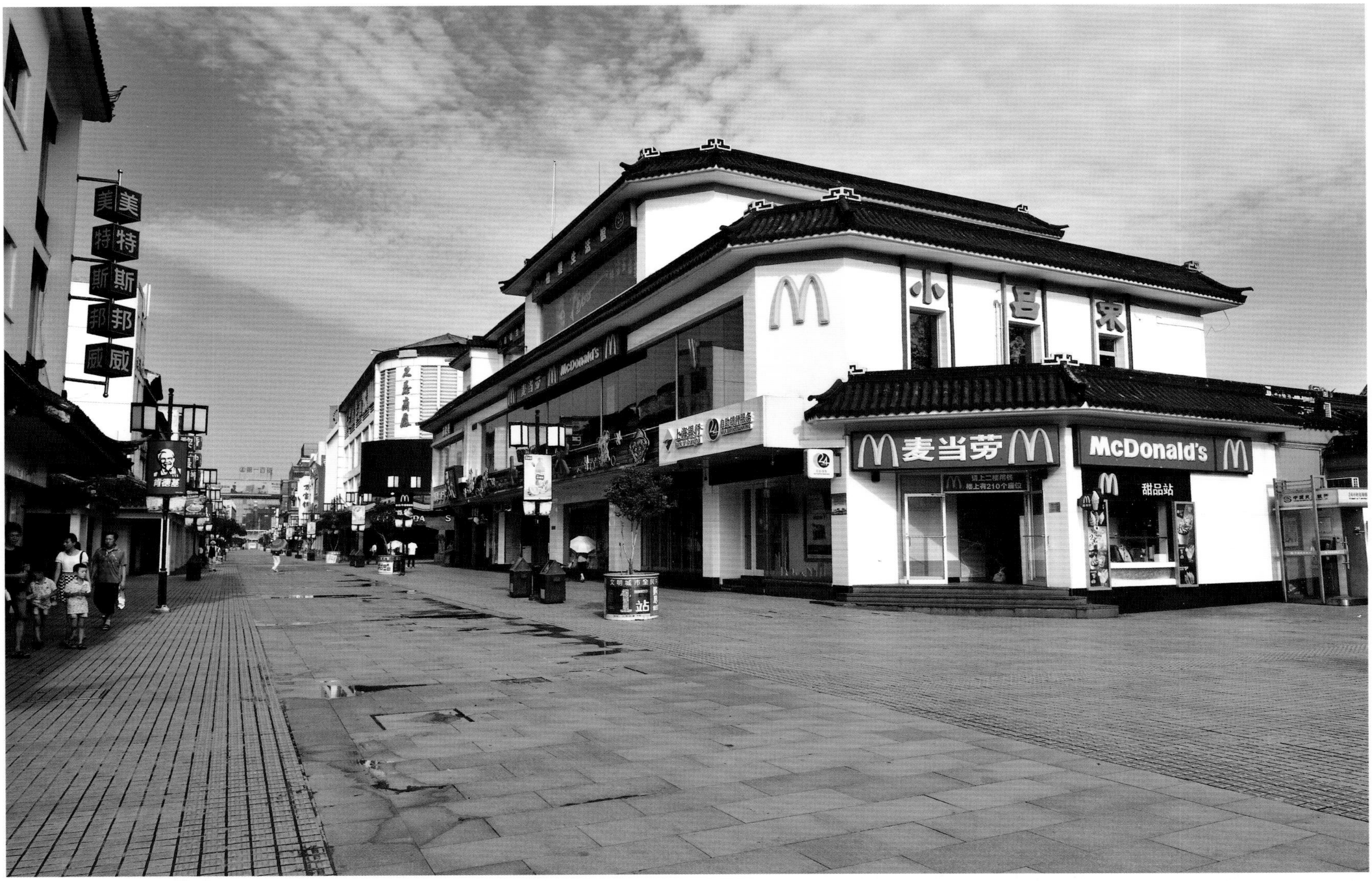
美特斯邦威
McDonald's
麦当劳
McDonald's
甜品站

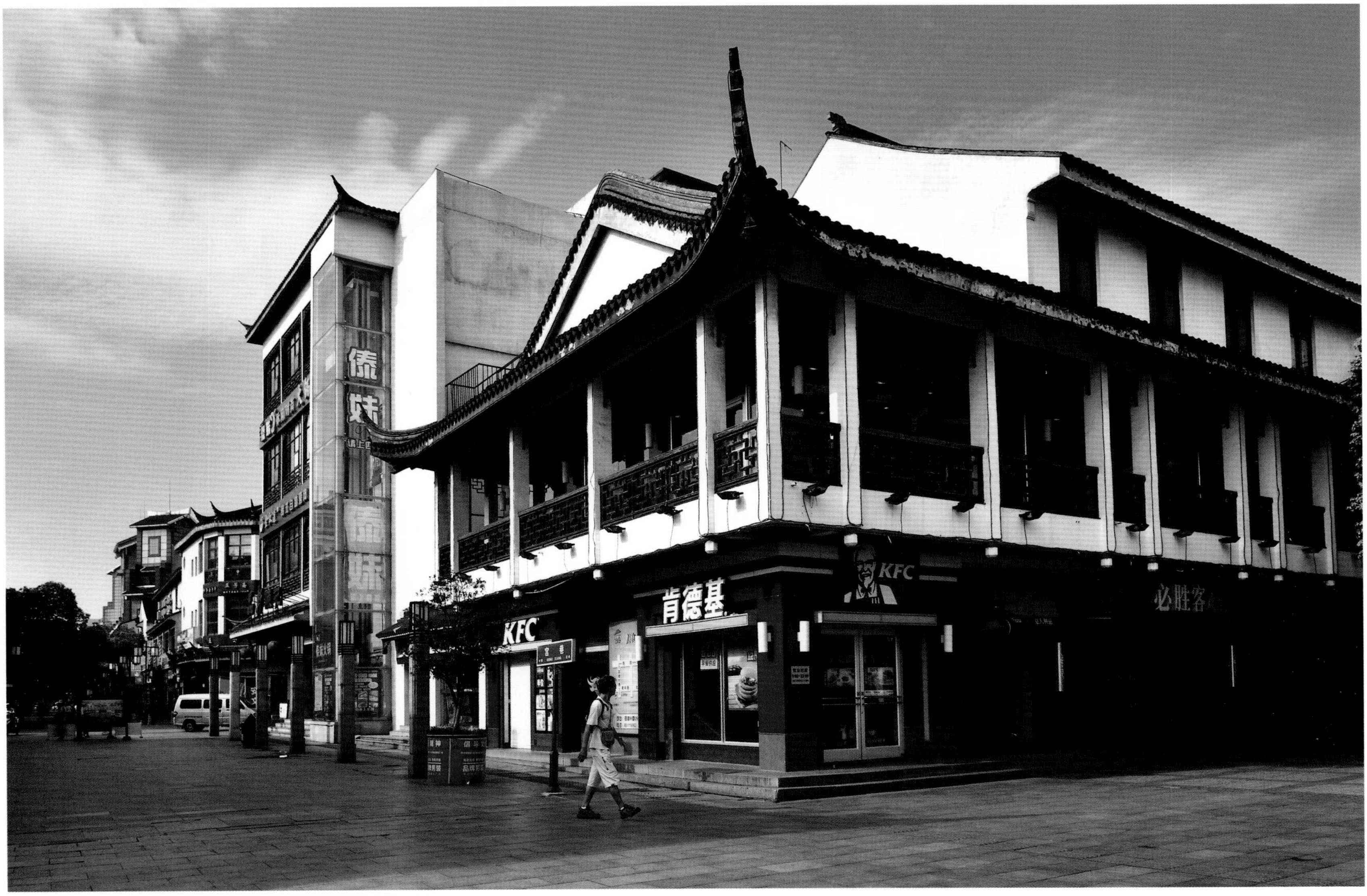
KFC
肯德基
KFC
必胜客

Commercial Activities

商业业态

Major Commercial Activities 主要商业业态

The Guanqian Street Area takes Guanqian Street as its major axis and core commercial zone, Taijian Lane as support axis and other lanes as radiate axes to form a comprehensive commercial street of multi-layer and multi-function considering culture, ecology and convenience. The types of commercial activities are various, including mix-box selling young female clothes, Qiantaixiang specializing in silk and satin, time honored pastry shops, traditional Chinese medicine stores, large scale shopping malls, book stores and so on. In addition, there are also shops , SPAs, nightclubs, KTV and others.

观前街街区以观前街为主轴和核心商业带，以太监弄为支轴，碧凤坊、宫巷、邵磨针巷为辐射商业轴，北局及玄妙观区域为支面，形成了轴轴与轴面相结合的，兼顾文化、生态和便利的多层次、多功能、复合型的商业街区。街内商业类型多样，有专营年轻女士服饰的mix-box，专售绫罗绸缎的乾泰祥；有百年品牌老字号特色糕点店黄天源、叶受和、稻香村；有中药馆雷允上；有大型商场豫园商厦、久泰商厦；有珠宝首饰类品牌店龙凤金店、万宝银楼、亨孚银楼。文化类的有光裕书厅（评弹）、大光明影楼、新华书店；除此之外，街内还有电器、电子数码类，足疗、SPA类、夜总会、KTV等诸多商业业态。

Clothes 服饰类

mix-box

mix-box, invested and monitored by British C&B Investment Holding Limited, is initiating advanced fashion and concept of continuous innovation. The target group of mix-box is fashionable women from 14 to 26 years old. Its products include fashionable accessories, cosmetics, cosmetic applicators and dolls.

mix–box

mix-box有英国C&B投资背景，并由British C&B Investment Holding Limited全球品牌监控，倡导超前时尚的主张和不断创新的理念，彻底混融新潮，抵制同化与平庸！mix-box的主要受众人群为14~26岁的女性潮人，经营品类有流行饰品、时尚小物、化妆品、化妆工具、玩偶五大类。

Qiantaixiang

Qiantaixiang was founded in 1863. It is one of the first batch of members of Suzhou trade association and also a time-honored brand. It is located at the middle part of Guanqian Street. Though it had been changed hand for several times in over one hundred years, its silk and satin business in Suzhou is very famous.

Now Qiantaixiang has made use of its advantage in Suzhou to establish special distributions in several places including Wuxi and Nanjing to expand its popularity and crediworthiness.

乾泰祥

乾泰祥绸布商店创办于清朝同治二年（1863年），创始人周以漠。该店是苏州商会的第一批成员，也是一个历史悠久的中华老字号。乾泰祥绸布商店位于观前街中段，是苏州绸布行业中的百年老店，百余年来虽数次易主，但“绫罗绡绉丝纱”的绸缎经营却在苏州城中非常有名。

现在，百年老店乾泰祥依靠经营苏州丝绸的优势，已经走出了苏州，在无锡市、南京夫子庙等处设立了丝绸特约经销点，扩大了其知名度和信誉度。

Food & Beverage 餐饮类

Caizhizhai

Caizhizhai is a Chinese time-honored candy store which was founded in 1870. It is a representative enterprise which is famous for its traditional Suzhou food. Products of Caizhizhai include three hundred kinds of five series including candy, cakes, roasted seeds and nuts, many of which are local specialties well-known at home and abroad.

Caizhizhai has a glorious history. Its products used to be royal tribute in the late Qing Dynasty, for its good effect on assisting medicine. In the 1950s, its products were used to serve foreign friends by Premier Zhou Enlai in Geneva Conference.

采芝斋

采芝斋是中国老字号糖果店，始创于清朝同治九年（1870年），是以生产苏式传统食品而著称的代表企业。采芝斋的产品包括苏式糖果、糕点、炒货、蜜饯、咸味点心等五大系列三百多个产品品种。采芝斋生产的许多产品是国内外知名的苏州特产，如酥糖、麻饼、粽子糖等。

采芝斋的历史辉煌。清末，采芝斋苏式糖果曾因苏州名医曹沧州献于慈禧助药，有很好的效果而成为“贡品”。20世纪50年代，周恩来总理在日内瓦国际会议上用采芝斋的糖果来招待国际友人而被誉为“国糖”。

Huang Tianyuan

Huang Tianyuan was founded in 1821 by Huang Qiting and later was transferred to Gu Guilin, a master worker in the shop. Huang Tianyuan only provided five-color dumplings, powder dumplings, lard cake, and yellow pine cake at the beginning. After several generations' hard work, its products have been developed into over 200 kinds and it provides only about 60 kinds everyday. Later it innovatively developed other kinds of gift type dumplings. Now Huang Tianyuan's products are sold at home and abroad by its branches.

黄天源

黄天源糕团店始创于清朝道光元年（1821年），创始人浙江慈溪人黄启庭，后转让给店中师傅顾桂林，因而黄天源糕团店实为顾姓店铺。黄天源糕团店中起初供应的品种仅有五色汤团、挂粉汤团、猪油糕、黄松糕等几种，后经过多代努力，黄天源糕团品种发展至200多种，每天供应60余种。后来又创新发展了其他礼品糕团，并从蛋糕裱花中得到启发，设计出了松鹤同春、龙凤呈祥、嫦娥奔月等口味美、造型美的新型糕团。现在的黄天源糕团通过分店行销海内外。

Ye Shouhe

Ye Shouhe food store was founded by Ye Hongnian in 1886. At the beginning, all its products were Suzhou style pastries. However, later its pastries were integrated Ningbo features, for its second and third managers were from Ningbo. It was merged into Suzhou pastry factory during the "Cultural Revolution". After the reform and opening-up, it has recovered its old brand to welcome tourists from home and abroad with a brand new appearance.

叶受和

苏州叶受和食品商店，原名叶受和茶食糖果店，创立于清朝光绪十二年（1886年），创始人叶鸿年。创业初期，叶受和的糕点、炒货、野味、糖果均属于苏式糕点，因其第二任、第三任经理均为宁波人，因而其后的糕点融入了宁波糕点的特色，拥有小方糕、云片糕、四色片糕等特色糕点品牌。“文革”期间曾并入苏州糕点厂，改革开放后，叶受和恢复老招牌，以崭新的面貌迎接海内外的游客。

Wufangzhai

Wufangzhai founded in 1854 is a time-honored brand. During the Republic of China, Wufangzhai zongzi came out. It developed rapidly, for its exquisite appearance, good-quality materials and workmanship, and unique taste. Now the parent company of Wufangzhai has become a private enterprise group integrating food, agriculture, house property, finance and trade.

五芳斋

五芳斋始创于1854年，是中华老字号，在民国时期，五芳斋粽子问世，因其外形别致，选料、制作考究，风味独特，发展迅速。到现在，五芳斋的母公司已然是一家以食品产业为核心，融食品、农业、房产、金融、贸易于一体的民营企业集团。

Traditional Chinese Medicine 中药类

Lei Yunshang

Lei Yunshang was a Chinese medicine store founded in 1734 and was originally named "Leisongfen Hall". It was famous for the founder, Lei yunshang's brilliant medical skill; therefore its name was gradually neglected and was called Lei Yunshang. The Lei Yunshang store at Guanqian Street now is a branch of Shanghai Lei Yunshang Pharmacy.

雷允上

雷允上始创于清朝雍正十二年（1734年），创始人雷大升（字允上），店铺原名"雷诵芬堂"，是一家中药店。因他以自己的字"允上"在店内挂牌坐堂行医，兼医术高明，治病有方，于是"雷允上医生"之名闻于苏州，后来人们称药店为"雷允上"，原来药店的名字却被世人遗忘，鲜为人知了。如今位于苏州观前街的雷允上药店是上海雷允上药业的分店之一。

Jewelry 珠宝类

Laofengxiang

Founded in 1848, Laofengxiang has a long history and profound culture. After more than 160 years development, Laofengxiang has become one of the "most valuable brands" leading enterprise in jewelry industry in China.

老凤祥

创建于1848年的中国民族品牌"老凤祥"有着悠久的历史和深厚的文化底蕴，历经160余年的发展，老凤祥集科工贸于一体、产供销于一身，发展成为中国珠宝首饰业"最有价值品牌"和龙头企业之一。

文化设施 Culture Facilities

Xuanmiao Temple 玄妙观

Founded in 276, Xuanmiao Temple is located at the prosperous downtown area of old Suzhou. It is facing Guanqian Street at its south. It is famous for its long history, grand buildings and numerous cultural relics. It is not only a scenic spot in Suzhou, but also one of the important Taoism temples in the country.

玄妙观创建于西晋咸宁二年（276年），坐落于苏州古城中心繁华的闹市区，南临观前街，面对宫巷，以其历史悠久、建筑宏伟、文物古迹众多而蜚声天下。它既是江南著名古观和苏州的一大名胜，也是全国重要的道教宫观之一。观内现有正山门、文昌殿、观音殿、雷尊殿、三清殿、四海亭、六合亭等建筑。

引导指示系统 Guidance & SignSystem

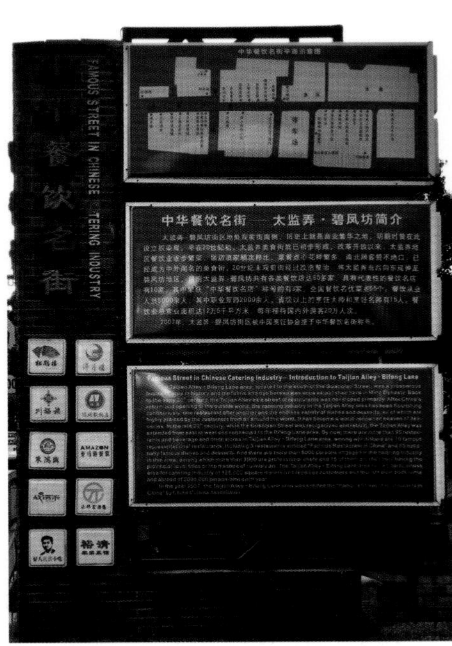

Dongguan Street, Yangzhou
扬州东关街

街区背景与定位
Street Background & Market Positioning

History 历史承袭

Dongguan Street is about 1,200 years old. After the Grand Canal opened up, this street gradually became a gathering place of most active trades and culture communication. Through over a thousand years, abundant historic sites and cultural relics are left in the street.

In Tang Dynasty, it was praised as "the top commercial port in southeast China", and the Dongguan Ancient Ferry was the most prosperous traffic communication centre in Yangzhou. Because of convenient and busy shipping, a prosperous street — Dongguan Street was created. Dongguan Street had become a famous commercial street since Tang Dynasty.

In Qing Dynasty, Yangzhou satlter, one of the four top itinerant traders of China, took Dongguan as the first choice for dwelling. In the late Qing Dynasty and early Republic of China, Yangzhou modern industry and commerce initiated, and Dongguan Street was the birthplace of numerous time-honored shops.

东关街约有1 200年历史。自京杭大运河开通后，这条外依运河、内联城区的通衢大道，逐步成为最活跃的商贸往来和文化交流集聚地。经过千年的积淀，街内留下了丰厚的历史遗存和人文古迹，堪称京杭大运河沿线城市中保存最完好的商业古街。

唐代，扬州赢得了“东南第一商埠”的美誉，有天下“扬一益二”之称，而利津古渡（即今天的东关古渡）是当时扬州最繁华的交通要冲。有了码头就有街市，舟楫的便利和漕运的繁忙，催化出一条商贸密集、人气兴旺的繁华古街——东关街。自唐以后，关东街一直是扬州著名的商业街。明嘉靖三十五年（公元1556）年，为防倭寇侵扰，扬州知府吴桂芳依附旧城东郭建外城，即从旧城东南角起，循运河向东，折到北，复折向西，到旧城东北角止，称为“新城”。建成的新城，以小秦淮河与旧城相连，东、南、北三面，设钞关挹江门、徐凝南便门、缺口通济门、东关利津门、便门便益门、广储镇淮门、天宁拱宸门7座城门。

清代，中国四大行商之一的扬州盐商，更把东关街作为居住的首选之地。清末民初，扬州近代工商业开始启蒙，东关街遂成为众多商业老字号的发祥地。

Location 区位特征

Dongguan Street extending from ancient canal at the east to Guoqing Road at the west is 1,122 meters long. It is not only the vital water and land communication line of Yangzhou, but also the center of business, handicraft industry and religious culture. The street is very bustling with all kinds of trades.

东关街东至古运河边，西至国庆路，全长1 122米，原街道路面为长条板石铺设。这里不仅是扬州水陆交通要道，而且是商业、手工业和宗教文化中心。街面上市井繁华，商家林立，行当俱全，生意兴隆，陆陈行、油米坊、鲜鱼行、八鲜行、瓜果行、竹木行近百家之多。

Market Positioning 市场定位

To highlight the featured positioning of Dongguan Street's four "theme area": first, "Wencuiyuan" centralizing Ge Garden highlights the refection of historic cultural elements such as salter culture and imperial art academy; second, "fitness relaxation workshop" centralizing Xiefuchun emphasizes fitness relaxation culture of Yangzhou; third, "gourmet relaxation workshop" centralizing Sanhesimei Pickle Factory highlights the food and snack of Huaiyang; fourth, "Song Dynasty Historic Site Square" connects Dongguan Street with the ancient canal.

突出东关街区的四大“主题区”特色定位：第一，以个园为中心的“文萃苑”，突出反映东关街盐商文化、扬州画院等历史文化要素；第二，以谢馥春为中心的“保健休闲坊”，突出反映扬州保健休闲文化；第三，以三和四美酱菜厂为中心的“美食休闲坊”，突出淮扬美食小吃，弘扬扬州的美食文化；第四，以城门遗址为中心的“宋城遗址广场”，将东关街区与古运河联系起来。

规划设计特色

Planning & Design Features

Street Planning 街区规划

After the city developing layout of "ancient town reveals cultural connotation, new area shows modern civilization", Dongguan Street as the core area of ancient town of Ming and Qing dynasties has become a pilot project. By using the German and Chinese technical cooperation project "eco-city planning and management" the city cooperates with UN-Habitat and GTZ.

On the basis of insisting drawing up the Dongguan Street protection planning from a high starting point, the city has entrusted departments like Southeast University to draw up several detailed plannings to provide protecting and developing Dongguan Street with scientific basis. In addition, it invited some famous professors, such as Liu Taige, Wu Liangyong, Wang Jinghui, Ruan Yishan and so on to strengthen the direction of Dongguan Street project.

确立“文化底蕴看古城、现代文明看新区、风景名胜看蜀冈、经济实力看沿江”的城市发展布局，东关街作为明清古城区的核心区域，成为“护其貌、美其颜、扬其韵、铸其魂”的试点工程。利用中德“生态城市规划与管理”技术合作项目，与联合国人居署、德国技术合作公司就古城保护理念、民居修复方案、资金筹措渠道等方面进行密切合作。

坚持高起点编制东关街保护规划，在老城区控制性详规、老城区12个街坊控制性详规的基础上，委托东南大学等单位先后编制了《东关街“扬州传统风情文化街”概念规划》《“双东”街区保护与整治引领规划》《“双东”街区“一点十片”详细规划》等，为东关历史街区保护、利用提供了科学依据。此外，还聘请了刘太格、吴良镛、王景慧、阮仪山等著名专家、学者为顾问，以加强对东关街保护整治工程的指导。

Street Design Features 街区设计特色

Its twist and deep lanes, and residences with black bricks congeal prosperity in the past and reflect the history of the city. Nineteen historic sites are protectively repaired; over three thousand old houses are renovated; environmental quality and public facilities are improved and folk arts and crafts like paper-cut are protected.

The general design keeps a comparably completed building groups of Ming and Qing dynasties and "fishbone" streets and lanes layout to inherit the traditional appearance of Ming and Qing dynasties. There are many important historic relics including two national level, two provincial level and twenty one city level culture relic protection units. This project forms an active space layout and presents the unique charm of a Jiangnan canal City.

巷道曲折幽深，民居青砖黛瓦，凝结着往昔繁华，折射出城市历史。在东关历史街区保护中重点对一批文化古迹进行保护性修缮，如个园南部住宅、逸圃等19处古迹；对危旧房屋进行分类整治，共整治3 000多户；完善基础设施，整修老街巷26条；提升环境质量，增加绿化、公厕、停车场等设施；保护剪纸等民间工艺。

整体设计保持了比较完整的明清建筑群及“鱼骨状”街巷体系，沿袭了明清时期的传统风貌特色。街内现有50多处名人故居、盐商大宅、寺庙园林、古树老井等重要历史遗存，其中国家级文保单位2处，省级文保单位2处，市级文保单位21处。形成了“河（运河）、城（城门）、街（东关街）”多元而充满活力的空间格局，体现了江南运河城市的独有风韵。

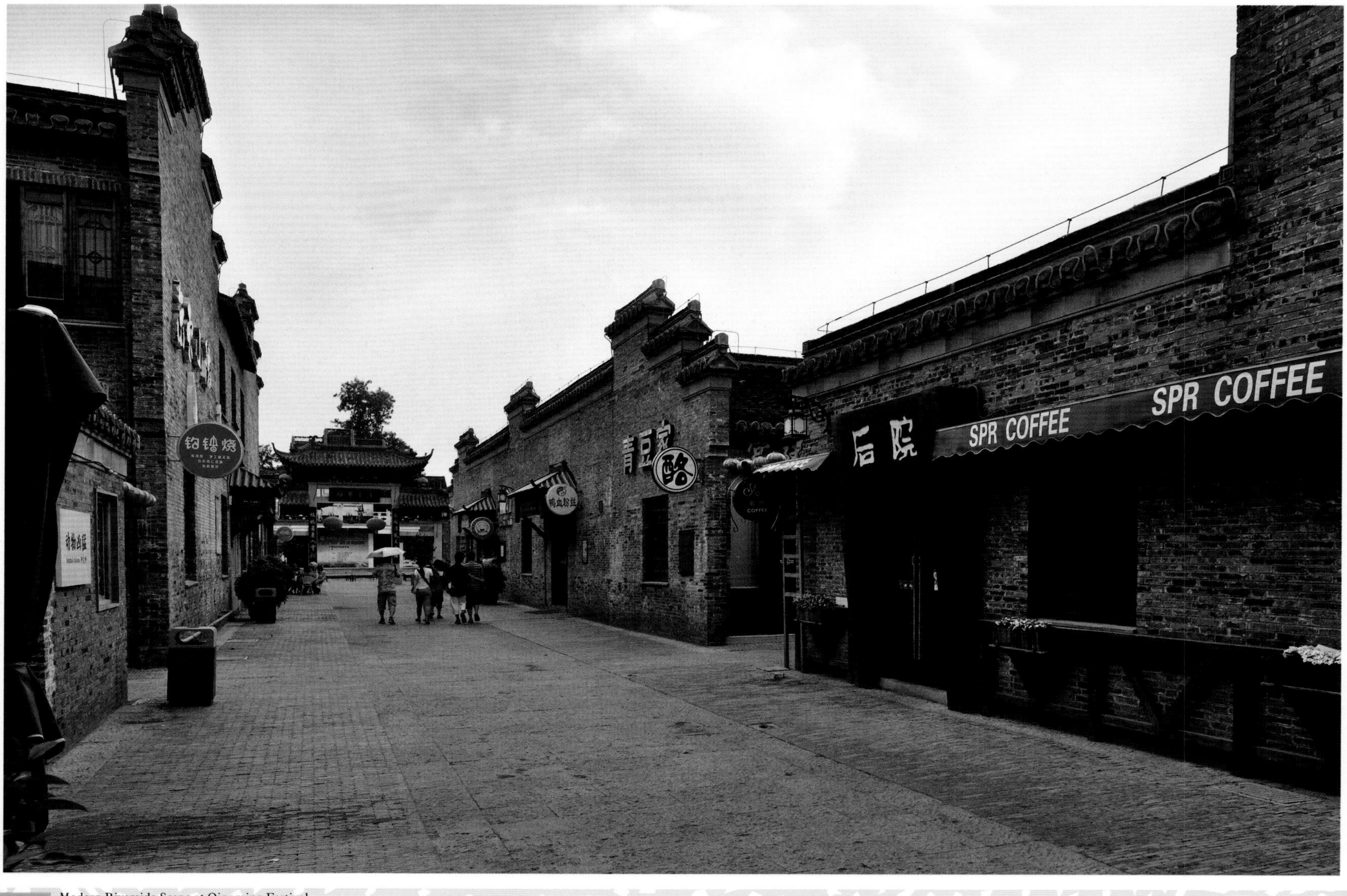

国家级旅游景区
AAAA

空调开放
本店有发票
鱼面 藕粉圆
藕 糯米藕
扬州炒饭 大煮干丝 各色卤菜 淮扬菜
特色面条 各色馄饨 各色藕粉 藕粉圆
营业时间

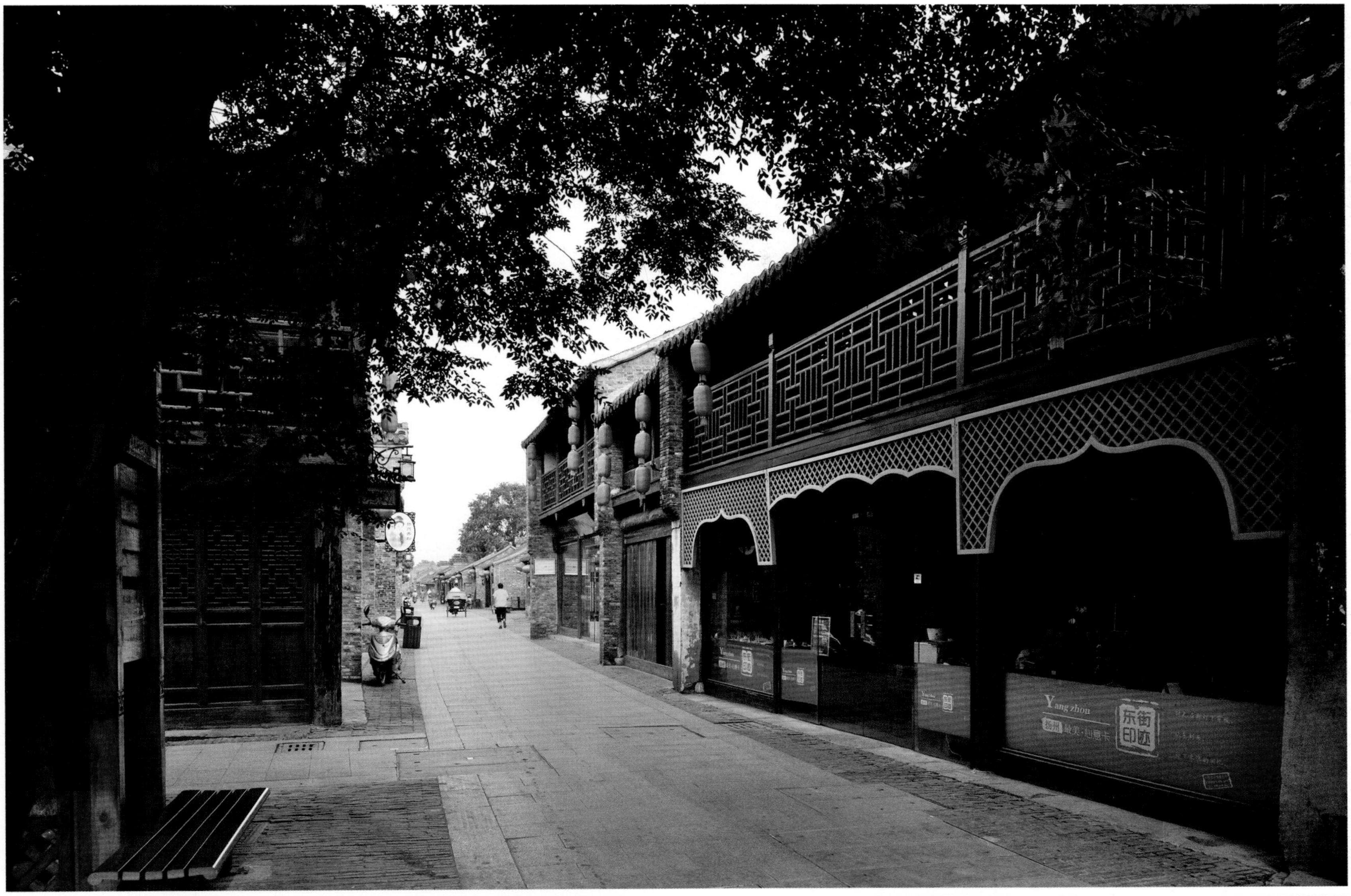

营业時間
10:00
22:00
原创品牌小店
SELF
LIFE
FAMILY

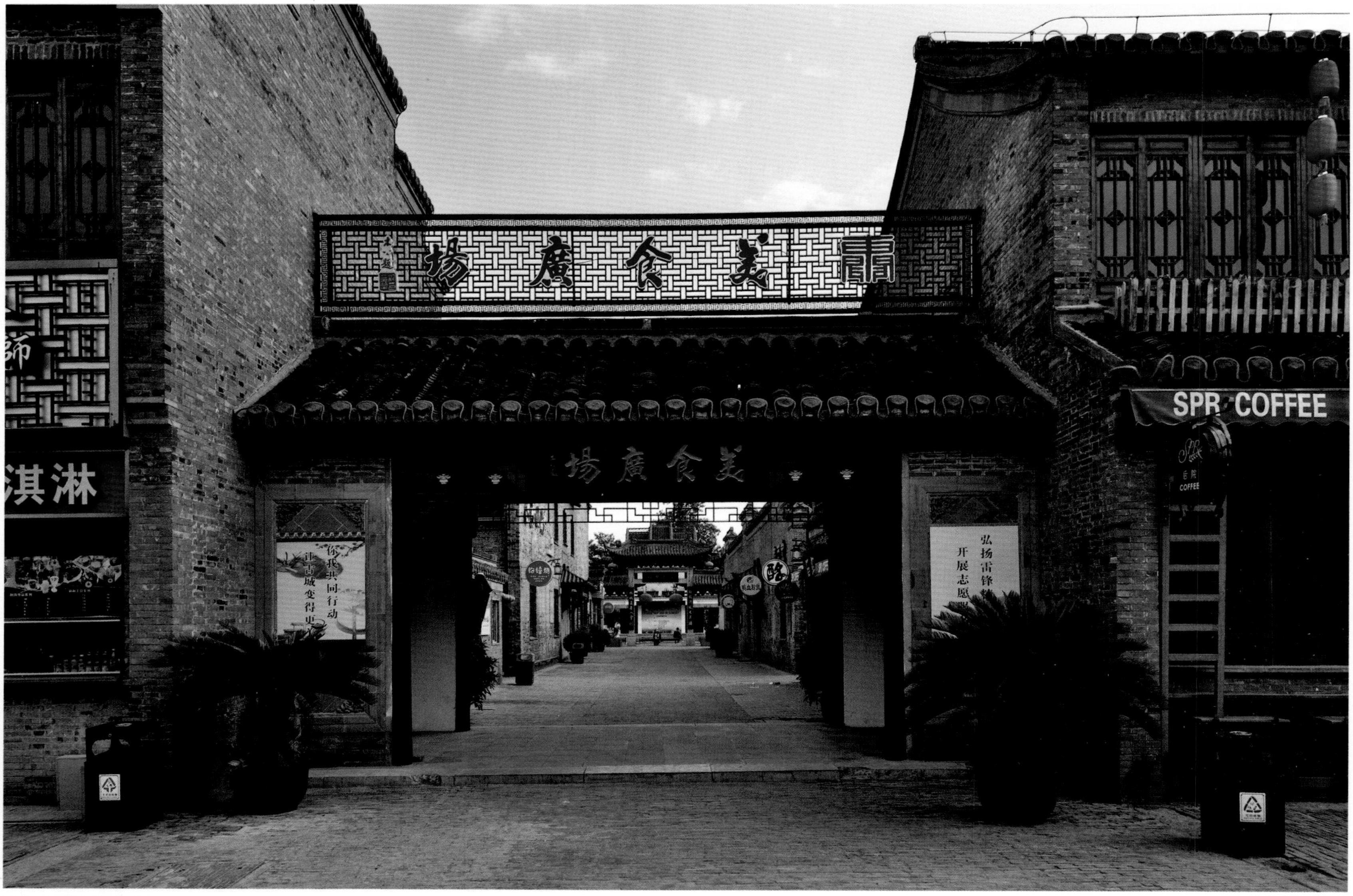
美食廣場
美食廣場
SPR COFFEE
淇淋

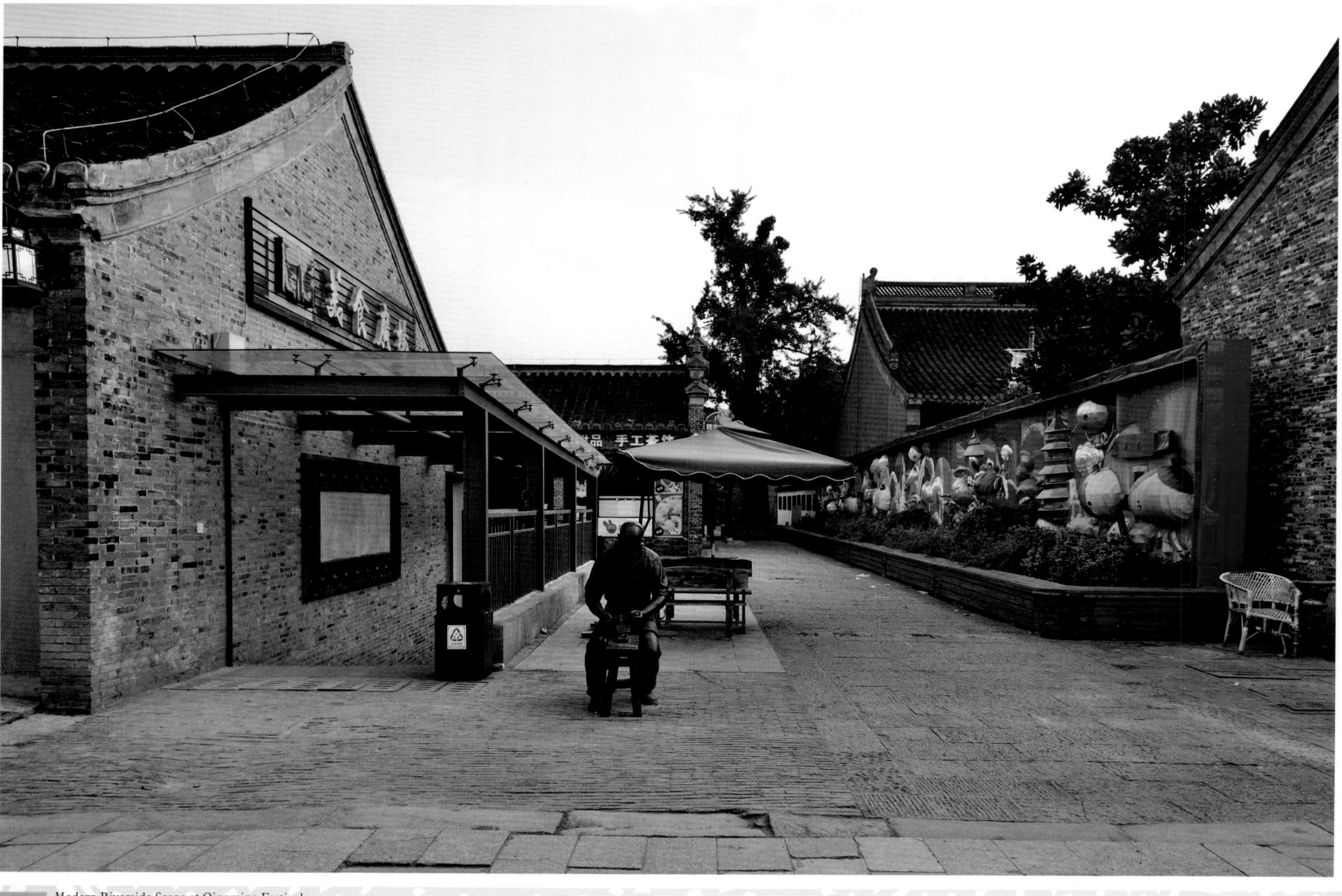
美食廣場

Major Commercial Activities 主要商业业态

Up to now, Dongguan Street is still an important place of business in Yangzhou and gathers handicrafts of local traditional features, featured snacks and time-honored shops, all together 232 individual businesses, among which are 72 handicraft industries, 24 restaurants and 136 tourist goods retailers including 19 traditional old and famous shops.

As for arts and crafts, there are paper-cut, sculptures, Guqin, Guzheng and so on; as for curio, there are over ten shops selling calligraphies and paintings, jade articles and lacquerwares; as for featured food, the street features Huaiyang dishes, matched with leisure non-traditional food to cater for young generation.

时至今日，东关街仍是扬州的商业重地，汇集了当地传统色彩浓厚的手工艺、特色小吃和商业老字号，全街共有个体工商户232家。其中，手工业72家、餐饮业24家、旅游商品经营户136家，包括19家传统商业老字号，如“百年老店绿杨春”、有“中华首妆”之称的谢馥春、久负盛名的传统名特产品“三和四美”酱菜、扬州文字记载中最早的药店“协茂大药房”以及“大清盐号”等。

在工艺美术类方面，有剪纸、雕刻品、古琴、古筝特色商品；在古玩类方面，包括字画、玉器、漆器等共有10多家；在特色餐饮类方面，以扬州美食为主打，主推淮扬菜，并在此基础上搭配引入休闲型的非传统餐饮，以符合年轻一代的口味。

Food & Beverage 餐饮类

Daqilin’ge

Daqilin’ge founded in 1901 is an old shop subordinated to Yangzhou Wuting Food Co., Ltd. For over one hundred years, and pastries of Daqilin’ge are favored by consumers for their elaborate craft, unique materials and graceful appearance. In order to adapt to the development of market, Wuting Food Co., Ltd. set up a professional operation and management mode to make Daqilin’ge a traditional pastry producer of the biggest scale of produce and sale in the central area of Jiangsu.

大麒麟阁

大麒麟阁食品店创办于1901年，是扬州五亭食品有限公司旗下的一家百年老店。近百年来，大麒麟糕点以工艺精细、配方独特、造型典雅而深受消费者青睐。店内品种多达数百种，每个品种都集色、香、味、形于一体。2003年3月法国总统希拉克访问扬州时，品尝了大麒麟阁的糕点后，赞不绝口，并特意购买了大麒麟阁的京果粉、麻饼等中式糕点带回法国。为了适应市场的发展需求，五亭食品有限公司成立了专业的经营管理模式，大麒麟阁从原有的手工作坊式生产走上标准化、产业化的发展模式，成为苏中地区产销规模最大的传统糕点生产基地。

Lvye Sticky Candy

Lvye Sticky candy is praised as a unique talent in Yangzhou. It tastes fragrant and sweet with sesames outside. It is semitransparent and flexible. It enjoyed a high reputation at home and abroad, but was lost, for wars. However, after hundred and thousand times of try, Yangzhou food manufactures eventually reproduce the traditional food again and add innovations to make it more delicious.

绿叶牛皮糖

绿叶牛皮糖号称“扬州一绝”。口味香甜，外层芝麻均匀，切面棕色光亮，呈半透明状，富有弹性，味香，细嚼不粘牙，在海内外享有盛誉，早在清朝乾隆、嘉庆年间就已面市扬州，后因战争频繁而失传于世。扬州食品生产厂家通过千百次的尝试，终于使传统产品得以重新问世，并使其在保持原有特色的基础上加以创新，使口感达到了弹性、韧性、柔软性三性一体的最佳状态。

Yuanxing Garden Huaiyang Cuisine

Huaiyang Cuisine appeared in Chunqiu period, developed in Sui and Tang dynasties and flourished in Ming and Qing dynasities. Its development was benefit from Emperor Yang of Sui's tour to Jiangdu which brought cooking of North China. Yuanxing Garden integrates many flavors and cooking skills to select materials cautiously, cook elaborately. Its dishes taste fresh and delicious, and are suitable for people from both the south and the north.

园兴园精致淮扬菜

淮扬菜与鲁菜、川菜、粤菜并称为中国四大菜系，始于春秋，兴于隋唐，盛于明清，素有“东南第一佳味，天下之至美”之美誉。该菜系的发展得益于隋炀帝下江都，带来了北方烹饪手法，融合江南本土鲜美的食材后得到极大发展。园兴园融合百家，选料严谨，制作精细，注重本味，讲究火工，清鲜平和，浓纯兼备，咸甜适中，南北皆宜。

Jiangnan Yipin

Featuring feudal official healthy dishes and Yangzhou salt dealer dishes, Jiangnan Yipin has become a gathering place of high-class Huaiyang cuisine, for its unique and strong Yangzhou flavor.

江南一品

以官府养生菜、扬州盐商菜为特色的“江南一品”餐饮会馆，因其独特、浓郁的扬州味，成为高档品牌淮扬菜的聚集地。会馆由淮扬菜大师王恒余主理，如果用极其精练的词汇来概括“江南一品”的特点，就是景美、馔珍、人秀。

Leisure 休闲类

Xiefuchun

It was founded in 1830 and is the first ancestor of cosmetic industry in China. Its traditional products include “incense, powder, and oil”. Its representative incense is made of natural materials and is produced by elaborate craft. It was selected as tribute in Qing Dynasty.

谢馥春

扬州谢馥春化妆品有限公司已有177年的历史传承，是中国化妆品业的始祖。谢馥春香粉铺于1830年创建开办，品种有香粉、藏香、棒香、香袋等产品。制作的香粉因形似鸭蛋而闻名于世，采取天然原料，经鲜花熏染，冰麝定香工艺精制而成，具有轻、红、白、香之特点，为清廷贡粉，百姓冠称“宫粉”。

Xiciju

Both of its decoration style and product's novelty are unique. It features flower sculptures, hand painted porcelains, and upscale censer made of special soil. The workmanship, painting and baking of its products are all elaborately designed.

惜瓷居

惜瓷居无论从装修风格上或是货品的新颖上，都别具一格。主打产品包括花雕刻、手绘瓷器（台灯、摆设、挂件、首饰等），还有用特别的匣钵土烧制而成的高档香炉，产品从做工、绘画、烧制都经过精心设计。

Lvyangchun Tea Shop

LvyangchunTea Shop, a prestigious old shop, was founded in Guangxu Period of Qing Dynasty. It has changed its name for three times. It has developed a famous tea named "Kuilongzhu", which enjoys a great prestige for its pure and fragrant.

绿杨春茶叶店

绿杨春茶叶店是久负盛名的百年老字号，开创于清光绪年间，初始店名为“吉泰茶叶店”，后更名为“景吉泰茶叶店”，并研制出了享有盛誉的“一江水煮三省茶”的“魁龙珠”名茶。1956年，景吉泰参加公私合营，与森泰等几家同业改组为“绿杨春茶叶店”。经营的茶叶以“汤清、味浓、入口芳香、回味无穷”的特点而享誉绿杨城。

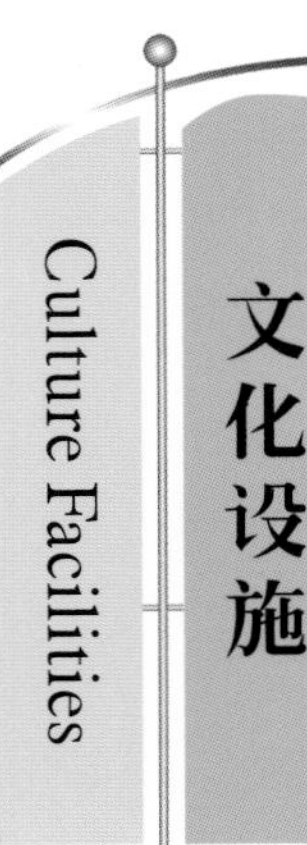

Ge Garden 个园

Ge Garden is a garden of a private residence. Its four-season rockeries are distinctive. It is praised as one of the four top gardens in China, and now it is a national level AAAA scenic spot.

个园是一处私家住宅园林。四季假山各具特色，以竹为名，以石为胜，被誉为中国四大名园之一，现为国家AAAA级景区。

Yipu Garden 逸圃

Yipu Garden was built by Li Hesheng in the early Republic of China. With houses at the west and garden at the east, it creates a poetic artistic conception.

逸圃在民国初年由钱业经纪人李鹤生所筑。逸圃西宅东园，整体呈现出“开门见山”“小中见大”的意境。

South Street Book House 街南书屋

South Street Book House is a relic of salt dealers — Ma Yueguan and Ma Yuelu's residential garden in Yongzheng and Qianlong period of Qing Dynasty. There are twelve scenes in the house, and the most famous one is a small and exquisite rock bought from Taihu.

街南书屋为清雍正、乾隆年间盐商马曰琯、马曰璐兄弟二人住宅园林遗址，因其在东关街南，故称“街南书屋”。街南书屋曾有12景，其中最负盛名的当数一块从太湖购来的小玲珑山石，因此，书屋又被称为“小玲珑山馆”。

Wudang Palace 武当行宫

Its original name was Zhenwu Temple. It was rebuilt by the county chief Chen Zhen in 1428. The three completed halls enable visitors to appreciate the original elegant demeanor of temples in Yangzhou.

武当行宫原名真武庙，明宣德三年（公元1428年）郡守陈贞重建，三座完整的大殿可让人原汁原味地领略扬州寺庙建筑的风采。

Nantang Street, Ningbo
宁波南塘老街

街区背景与定位 Street Background & Market Positioning

History 历史承袭

Nantang Street is located outside south ancient city gate of Ningbo. It used to be a business gathering place of old Ningbo and is one of the eight historic streets in Ningbo.

南塘老街位于宁波古城南门外，曾经是旧宁波商贸文化聚集地的“南门三市”，位列宁波八大历史街区之一，同时也是宁波市“紫线规划”保护区域。

Location 区位特征

The whole project extends from Xindian Road at the south to Yinjiang'an Road at the north, adjacent to Nantang River at the west and Yinfeng Road at the east. Its building area is 80,000 m² and is developed in two phases.

The first phase is adjacent to Zuguanshan Road at the south, Yinjian'an Road at the north. Its site area is about 49 500 m² and the total building area is about 38,000 m².

南塘老街整体项目南至新典路，北临尹江岸路，西起南塘河，东到规划路、鄞奉路，总建筑面积约80 000平方米，分两期实施开发。
一期南至祖关山路，北临尹江岸路，西起南塘河，东到规划路，占地约49 500平方米，总建筑面积约38 000平方米。

Market Positioning 市场定位

Through the deep exploring of Ningbo spirit and custom culture of Nantang River historic street and on the basis of protecting and restoring buildings layout and natural landscape of Nantang Street, it is positioned to be a historic cultural commercial street showing the features of Ningbo.

通过对南塘河历史街区内保留的宁波城市精神与民俗文化的深度挖掘，在保护与还原南塘老街建筑格局与自然景观的基础上，打造出集历史古迹、旅游观光、文化休闲、宁波老字号、宁波名优特产、民俗特色于一体，展现宁波江南水乡城市特征的历史文化特色商业街区。

Street Planning 街区规划

By applying various methods like protecting, repairing, reconstructing and so on to preserve and present historic appearance and cultural customs of the street in a largest extend.

According to the planning, the street can be divided into five functional areas:

Culture exhibition area — centralize historic buildings like Yuan Muzhi's former residence to build a cinema museum of most impact in Jiangnan.

Traditional retail area — mainly experience studios of traditional handicraft, local specialty selling and specialty catering.

Restaurant & bar entertainment area — mainly sightseeing, bars, business and hotels.

Institution office club area — mainly small size business offices.

Comprehensive commercial residence area — complex of business, cultural entertainment and part of residential function.

采用“保护、修缮、改建、迁建”等多种方法，最大程度上保留并体现出原有街区的历史风貌和人文风情，恢复和构建原有的老街市店铺、石板路、马头墙、河埠头等建筑空间意象。在保留与传承悠久甬商文化的同时，重现极富宁波本地特色的沿街店铺与兴旺市集，旨在发掘老宁波人的城市记忆，打造原汁原味的宁波特色历史文化街区。

按照规划，南塘河历史街区将分为以下5大功能区。

文化展示区——以袁牧之故居等历史建筑为核心，打造成江南地区具有影响力的电影博物馆；以袁氏居宅建筑群为核心，构建一个地方戏曲、地方语言爱好者学习交流的场所。

传统商贸零售区——以传统手工艺体验式工坊、土特产经营、特色餐饮为主。

餐厅酒吧娱乐区——以旅游观光、酒吧餐饮、商业、旅馆为主。

机构办公会所区——以小型机构商务办公为主。

商贸住宅综合区——商业、商务、文化娱乐及部分居住功能混合。

Street Design Features 街区设计特色

Inheriting the essence of Ningbo culture

Themed by "window of Ningbo traditional culture, features of Nantang River, area of Yuan family culture", the over 500 meters long Jiangnan traditional street is completely preserved to show business trade and daily life of Ningbo of one hundred years. There is a cultural relic protection unit and 6 cultural relic protection sites.

Reproducing "layout of one river, one road and one market"

Keeping the interdependent layout of Nanjiao Road, Nantang River and old Nanmensan Market according to its original city texture to strengthen the "fishbone structure" of the streets and lanes. new streets are integrated with old streets when protecting the historic appearance of the old building groups.

Import diversified leisure activities and featured retails

On the basis of inheriting historic building and culture, by fully using the advantage of location, diversified leisure activities and featured retails are imported to the street to add cultural business vitality. Following the definition of "yesterday, today and tomorrow", the first phase plans three theme restaurant retail areas to gather and present local food and food from over the country, and to show the pleased and relaxed mood of modern life.

秉承宁波文化之本

以"宁波传统文化之窗口，南塘河江南水乡之特色，袁氏文化之区域"为主题，通过完整保留500多米的江南传统街巷，再现百年来宁波人经商交易、日常生活的历史场景。街内拥有文保单位1处，文保点6处，并在设计中成功地将老街的历史神韵、建筑特色以及名人文化完美融入街区布局，实现"历史融入生活，商业融入古迹"。

重现"一河一街一市格局"

按照原有街区的城市肌理，保持了南郊路、南塘河、老南门三市这"一街一河一市"相依相辅的整体格局，强化脉络清晰的"鱼骨形"街巷结构，保护了古建筑群的历史风貌，并在新的街巷与老街巷的衔接中实现协调，赋予历史街区全新的生命力。

引入多样化的休闲业态及特色零售

充分结合南塘河历史街区优越的城市地理位置，通过对街区内历史建筑与文化印记的保护和传承，在结合宁波传统历史餐饮及酒楼品牌的基础上，引入多样化的休闲业态及特色零售，使古老街区再度绽放出独特的文化商业活力；遵循"昨天——今天——明天"的脉络定义，在一期项目中规划出三大主题的核心餐饮零售休闲区，分别将汇集和呈现宁波本地餐饮、各地美食与生活休闲餐饮以及融合现代"食"尚潮流的全新业态；多层次、多角度地演绎出现代生活的惬意休闲情调。

粥
家
莊

乾玉坊
QIAN YU FANG

Major Commercial Activities 主要商业业态

The street focally promotes traditional cultural food, and complementarily develops featured food and beverage and prevailing leisure entertainment. Here assembles dozens of shops of local tradition food and snacks including some time-honored shops. Relying on time-honored shops, Nantang Street hold various commercial market and traditional cultural activities to recreate the prosperity of Nanmensan Market.

After completing the first phase, there are totally over twenty shops opened. Apart from time honored restaurants of traditional food, numerous famous traditional snacks and restaurants from eleven counties of Ningbo are selected into the street to show the familiar taste in Ningbo people's memory and elaborately create a leisure eating experience of most local features for citizens.

街区内重点推广传统文化餐饮，兼顾发展特色餐饮和时尚休闲娱乐，汇聚了数十家来自宁海、象山、奉化、余化等宁波本土的传统特色餐饮和美食小吃，赵大有、草湖食品等一批宁波“老字号”企业在此设店。南塘老街将依托老字号商家，举办各种商贸集市和传统文化活动，使南门三市“夹道商铺，鳞次栉比，一如江东”的繁华在此重生。

一期街区开街后，共计有20多家店铺率先营业。不仅引入了传统餐饮老字号，更甄选宁波11县市区的众多知名传统特色小吃、餐饮和新派宁波菜，如宁海五丰堂、余姚黄鱼面、慈城四季香年糕等30余家商铺，共同演绎出属于宁波人记忆中熟稔的味道，融合甬上人文文化与“食”文化的多元特色，为现代宁波人精心打造出极具本土风情的餐饮休闲新体验。

Meilong Village

Its name is from a Beijing opera. It had been just a shop front at Weihai Road and provided Huaiyang snacks originally. After it was bought by Li Bolong, it was moved to Nanjing Road where it is located right now. It integrates Sichuan favor into Huaiyang dishes to form Shanghai style dishes.

梅龙镇

店名取自京戏《游龙戏凤》，最初在威海路只有一间店面，供应肉汤包等淮扬小吃。后由李伯龙买下，迁至南京路现址，并请名媛吴湄任经理，以淮扬名馔为号。将川味入扬，形成川扬合称的海派菜。店内海派名肴有龙园豆腐、芹菜鹌鹑丝、干烧鲫鱼、干烧桂鱼、干烧明虾、茉莉花鱿鱼卷、龙凤肉等，其中梅龙镇鸡口味别具一格。

Ningbo Yifu

It is one of the biggest shops of non-staple food in the 1980s in the country. It is a unforgettable childhood memory of young generation in Ningbo. Now Ningbo Yifu has become more characteristic, more humanized and more unique. Its products have also become more abundant including famous local specialty and leisure top grade non-staple food showing the feature and style of Nantang Street.

宁波一副

宁波一副是20世纪80年代全国最大的副食品商店之一，陪伴老宁波走过了峥嵘岁月，更是年轻一辈宁波人难忘的儿时回忆。如今的宁波一副更具特色化、人性化和个性化，内容也更加丰富，包括了名优特产、休闲及高档副食品，彰显了南塘老街作为宁波新文化窗口、旅游名片的风貌。以旗舰店、连锁店、社区店的销售网络形式，全面辐射宁波以及周边地区乃至全国市场。

Quanfeng Ji

It is said that when candidates for imperial examination got together in Zhejiang during Qing Dynasty, the owner of Quanfeng Ji in order to attract customers, pickled and then deep fried chicken which became famous. As an old brand of over one hundred years, Quanfeng Ji adopts selected materials, offers superior services. its chicken is extremely delicious.

全丰记

相传在清朝年间浙江举行考试时，各地考生齐聚，店主为了招揽客人，经“臭豆腐”做法的启发，将鸡肉腌制入味后油炸，于是外皮酥脆、肉多汁香的鸡排便和臭豆腐一样声名远播了。作为一个百年品牌，全丰记选材优质，服务精良，经过秘料腌制后的黄金鸡排外皮酥脆、多汁入口，绝对挑战人的味蕾。

Grand Mansion

Grand Mansion Restaurant Management Co., Ltd. has a modern restaurant management group of completed innovative development, enterprise operation and quality managing system. It mainly provides Ningbo dishes, Cantonese dishes and seafood.

大宅门

大宅门餐饮管理有限公司隶属大宅门餐饮集团直属企业，具有完善的创新开发、企业运营、质量管理体系和现代化餐饮管理团队，店内主营时尚宁波菜、粤菜、东海野生海鲜、南海海鲜、进口海鲜等。

Dumpling Shop

It wraps various delicious stuffing with thin dumpling wrappers and then cooks the dumpling to be glittering and translucent. Its dumpling with condiments is very delicious. Its classic stuffing includes chive with eggs, pork with cabbage, carrot with beef and so on.

饺子铺

用薄薄的饺子皮包裹各种美味的馅料，用水煮至晶莹剔透，蘸着佐料，回味无穷。店内比较经典的饺子馅，有韭菜鸡蛋、猪肉白菜、胡萝卜牛肉、素三鲜等，里面馅料很足，一口咬下去，汤汁瞬间满足味蕾。

Old Courtyard Restaurant

It features home cooking of Ningbo flavor. In the courtyard, customers, sitting around a small table with cattail leaf fans fanning, wine, dishes and talking to each other, can relax themselves and recall the "courtyard feeling" of old days.

老院子餐厅

餐厅主打宁波风味家常菜。老院之内，天井之下，左邻右里，一张小桌，几张小凳，蒲扇摇摇，喝个小酒，夹口小菜，聊个家常，感受久违的亲切，放松心情，不谈工作，在这里，可以回味从前的“老院味”。

South Gate Yuan Family Building Group 南门袁氏古建筑群

This building group was built in Qing Dynasty and composed of No.185 at Nanjiao Road, No.1 wing-room at Huici Lane, Yuan Family Ancestral Hall, Chongzhi Primary School and Yuan Family Hut. Its scale is grand and it is preferably preserved.

该建筑群为清代建筑，由坐西朝东的南郊路185号、惠赐巷1号偏屋、袁氏宗祠、崇志小学、袁氏家庵组成。此建筑群规模宏大，保存较好，是研究近代建筑的实物佐证。

Tongmao Ji 同茂记

It is a building of the Republic of China style with a total site area of 468 square meters. It is composed of a shop, a factory and opposite rooms.

同茂记为民国时期的建筑，总占地面积468平方米，由店铺、工场、倒座组成。店铺坐东朝西，重檐硬山顶结构，单开间；工场坐西朝东，主楼重檐硬山顶结构，五开间，六柱七檩；倒座单檐斜坡顶，三开间，二柱三檩。

引导指示系统

Guidance & Sign System

Sanfang Qixiang, Fuzhou
福州三坊七巷

街区背景与定位

Street Background & Market Positioning

History 历史承袭

After "An-shi Disturbances", refugees moved south to inhabit in Fuzhou, thus a dwelling region of literati and officialdom was formed. Dwellers in Tang Dynasty built a neat and orderly "new village" along the city axis, Nan Street. And then, a group of lanes was developed in a shape of "非". After hundreds years , its layout was formed in Ming and Qing dynasties and has became a symbol building of Fuzhou.

In December, 2006, the restoration work of Waterside Pavilion Stage officially began, which marked that the relic protection project of Sanfang Qixiang was in full swing. The relic protection project was basically completed in 2010. The ancient layout of this area is basically preserved. It is the only one "neighborhood system living fossil" in China's cities.

唐代"安史之乱"之后，南迁避难的人们聚居福州，形成了一个以士大夫阶层、文化人为主要居民的街区。唐代先民先是沿着城市的轴线——南街，建起了一组排列工整的"新村"。然后，再隔一条南后街，向西发展，建起一组坊巷，成为以南后街为中心轴线的"非"字形结构街区。经过千百年的风雨变迁，大多于宋代定下坊巷之名，于明清时期形成今天的建筑格局，成为福州历史文化名城的标志性建筑。

2006年12月，水榭戏台修复工程正式动工，标志着三坊七巷文保单位修复工程全面展开。2007年，福州市政府颁布《福州市三坊七巷历史文化街区古建筑搬迁修复保护办法》，整个工程总投资40亿元，全面修复159座明清古建筑，2010年基本完成。古老的坊巷格局至今基本保留完整，是中国都市仅存的一块"里坊制度活化石"。

Location 区位特征

Sanfang Qixiang is located at the center of the city. It is next to North Bayiqi Road at the east, Tonghu Road at the west, and adjacent to Yangqiao Road at the north and Dajibi Lane at the south. The west-ward three zones are called "Fang", and the east-ward seven lanes are called "Xiang".

三坊七巷地处市中心，东临八一七北路，西靠通湖路，北接杨桥路，南达吉庇巷、光禄坊，占地约40公顷，现有居民3 678户，人口14 000余人。向西三片称"坊"，向东七条称"巷"，自北向南依次为"三坊"——衣锦坊、文儒坊、光禄坊，"七巷"——杨桥巷、郎官巷、安民巷、黄巷、塔巷、宫巷、吉庇巷。

Market Positioning 市场定位

To create a tourist pedestrian street featuring leisure, restaurant and entertainment by using cultural resources and to create a historic cultural featured old street by its cultural relics and historic buildings.

利用文化资源，打造以休闲、餐饮、娱乐为主的旅游步行街；围绕"百年街肆的里坊遗址、历史风貌的建筑部落、闽都风韵的名人故居"，打造历史文化特色古街。

Street Planning 街区规划

Firstly, three exhibition zones are opened up: "special project exhibition zone" mainly showing genetic relationship culture of Fujian and Taiwan, "first comprehensive exhibition zone" showing culture of Lin family and shipping culture, and "second comprehensive exhibition zone" showing relative culture of men of literature and writing.

Secondly, different function zones are constructed according to the structure of "one river, one node, two streets and two zones". One river means the waterside leisure area of Antai River; one node means the "Guanglu Stage"; two streets are Nanhou Street traditional featured commercial area and Nan Street commercial development area; and two zones mean two concentrating exhibition zones with numerous cultural relics and historic buildings.

首先，开辟三个展示区："一个专项展示区"，即"七巷"中自郎官巷至黄巷之间的展示区，主要展示闽台亲缘关系文化；"第一综合展示区"为黄巷至吉庇路之间的展示区，主要展示林氏价值的各种文化和船政文化；"第二综合展示区"即"三坊"中各展示点有机结合形成的展示区，主要展示文人墨客的相关文化。

其次，借助"一水一节点两街两片区"的结构，构建不同功能区："一水"即安泰河的滨水休闲风情带；"一节点"是拟将光禄坊的"光禄吟台"作为三坊七巷南节点；"两街"为南后街传统特色商业带和南街商业更新发展带；"两片区"是结合北入口，将郎官巷、塔巷及水榭戏台部分作为旅游集中展示区。宫巷至安民巷有大量文物保护单位和历史保护建筑，作为博物馆建筑集中区；文儒坊南北段两侧作为会馆会所片区，以创意和休闲为主。

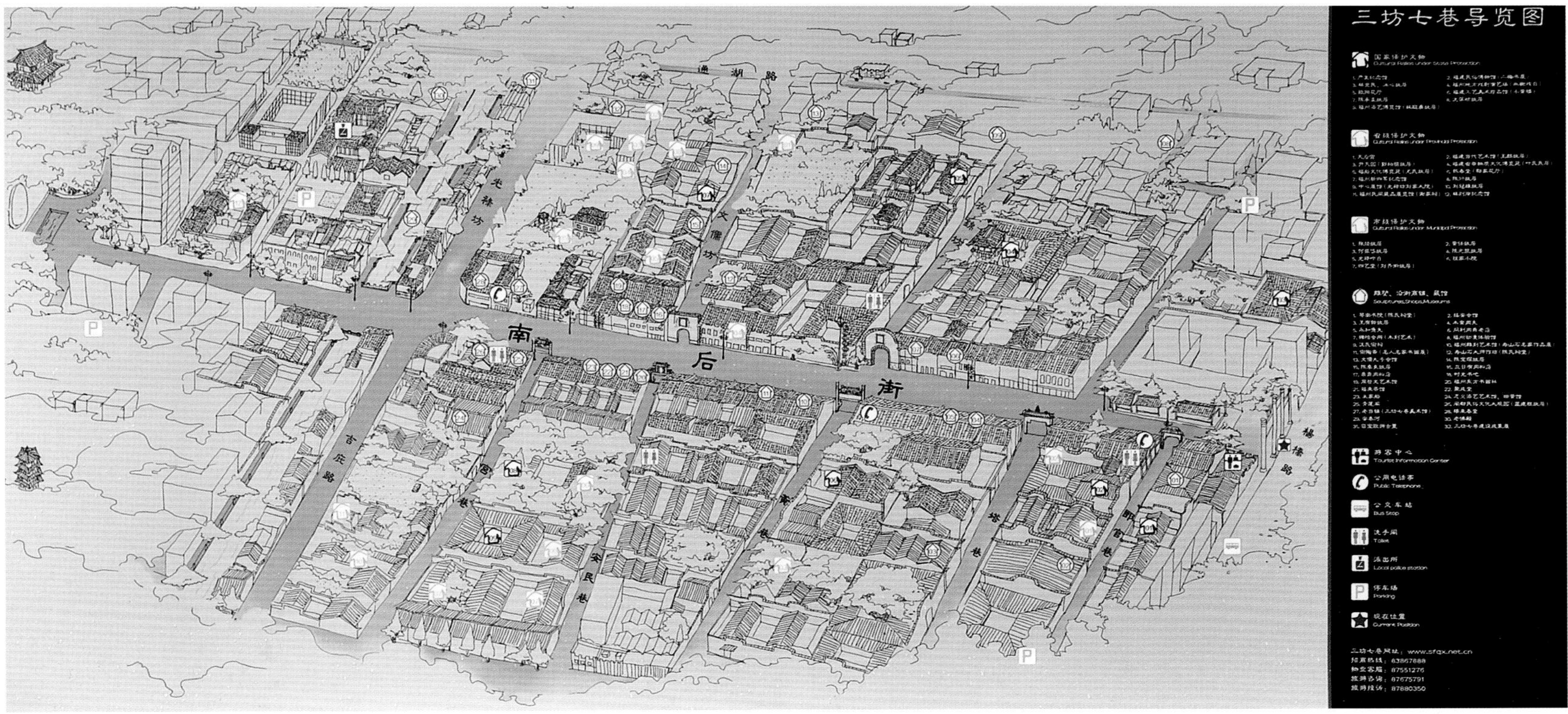

Street Design Features 街区设计特色

In this area, lanes and alleys are across each other, floor is paved with stones, and buildings are traditional houses.

Its architecture style has an ancient Fujian dwelling feature: strict layout, connecting courtyards, axial symmetry, wooden structure, saddle-shape fire walls, and dragons, phoenixes, flowers and bird patterns. The windows of principal rooms and rear rooms are double-layer connecting windows with the lower layer fixed and the upper layer operable. The doors of principal rooms are mostly four-door, and are decorated with rich patterns.

As for the space, the main hall at the axis works with other corridors, pavilions to form a high-low, active and changeable space layout. Halls are generally open style, which integrate with courtyard. In order to make the hall tall and spacious, columns applied thick and high-quality hardwood to reduce the number of columns.

Apart from distinct layout, Sanfang Qixiang has unique walls, carvings and doors.

Saddle-shape walls: they inherit the construction tradition of the late Tang Dynasty. They are generally symmetrical with clay sculptures and colored paintings at the heads formed a unique ancient Fuzhou residential appearance.

Carvings: columns in common dwellings are simple, yet their doors and windows are with elaborated carvings. The window decorations have rich types including the straight which have delicate density, the curve which are dynamic and the mixed which are changeable.

Doors: there are two kinds. One is right at the middle of front yard and is composed of stone frame in rectangle shape; the other is with saddle-shape walls at its two sides to form a big arch, such as doors of Shen Baozhen's, Chen Chengxi's and Lin Congyi's former residences.

街区内，坊巷纵横、石板铺地、白墙瓦屋、曲线山墙；不少还缀以亭、台、楼、阁、花草、假山，融人文、自然景观于一体。

在建筑风格上，具有闽越古城的民居特色：布局严谨，院落相连，中轴对称，以木结构承重，宅院四周或左右围有土筑的马鞍形风火墙，有的墙峰饰以飞龙飞凤、花鸟鱼虫及人物风景，具有浓郁的地方特色。正房、后房窗以双层通长排窗为多，底层为固定式，上层为撑开式或双开式。正房的主门朝大厅敞廊，多为四开式，门上雕有丰富的图案花饰，以增添大厅的气派之感。

在空间营造上，中轴线上的主厅堂与其他廊、榭等建筑形成高低错落、活泼而又极富变化的空间格局。厅堂一般为开敞式，与天井融为一体。为了使厅堂显得高大、宽敞、开放，一般在廊轩的处理上着力，采用粗大而长的优质硬木材，并用减柱造的办法，使厅堂前无任何障碍，这在北方建筑及其他南方建筑中，都极少见到。

三坊七巷除了在布局结构上与众不同之外，在围墙、雕饰、门面上也很有特色。

马鞍墙：沿袭唐末分段筑墙传统，墙体随着木屋架的起伏做流线型，翘角伸出宅外，状似马鞍，俗称马鞍墙。一般是两侧对称，墙头和翘角皆泥塑彩绘，形成了福州古代民居独特的墙头风貌。

雕饰：普通民居梁柱简洁朴实，而在门、窗扇雕饰上则煞费苦心。窗饰的类型特别丰富，有直线型、曲线型、混合型等，直线型疏密有致，曲线型富有动感，混合型变化多端，且各有吉祥寓意，充分显示了福建民间工匠的高超技艺。

建筑门：一种是在前院墙正中，由石框构成、与墙同一平面的矩形门；另一种则是两侧马鞍墙延伸做飞起的牌堵，马鞍墙夹着两面坡的屋盖形成较大的楼，像沈葆桢故居、陈承裘故居、林聪彝故居等都是这种门楼。

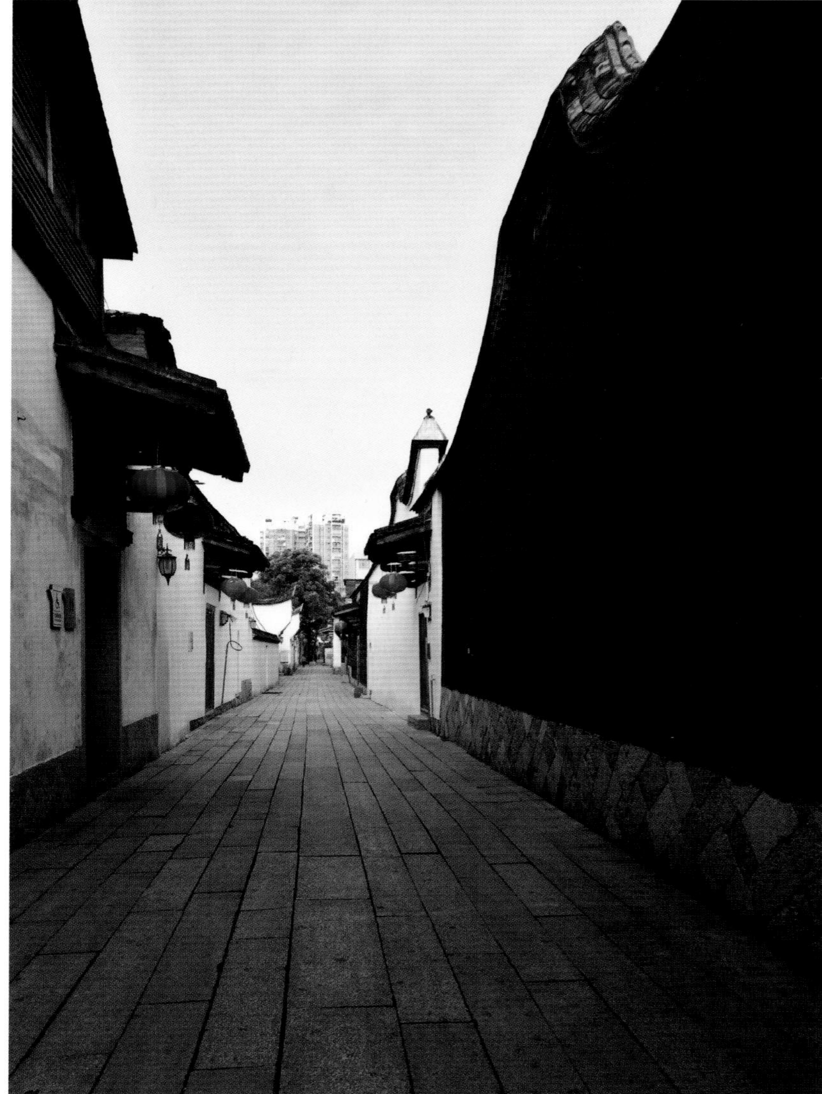

郎官巷

宫巷

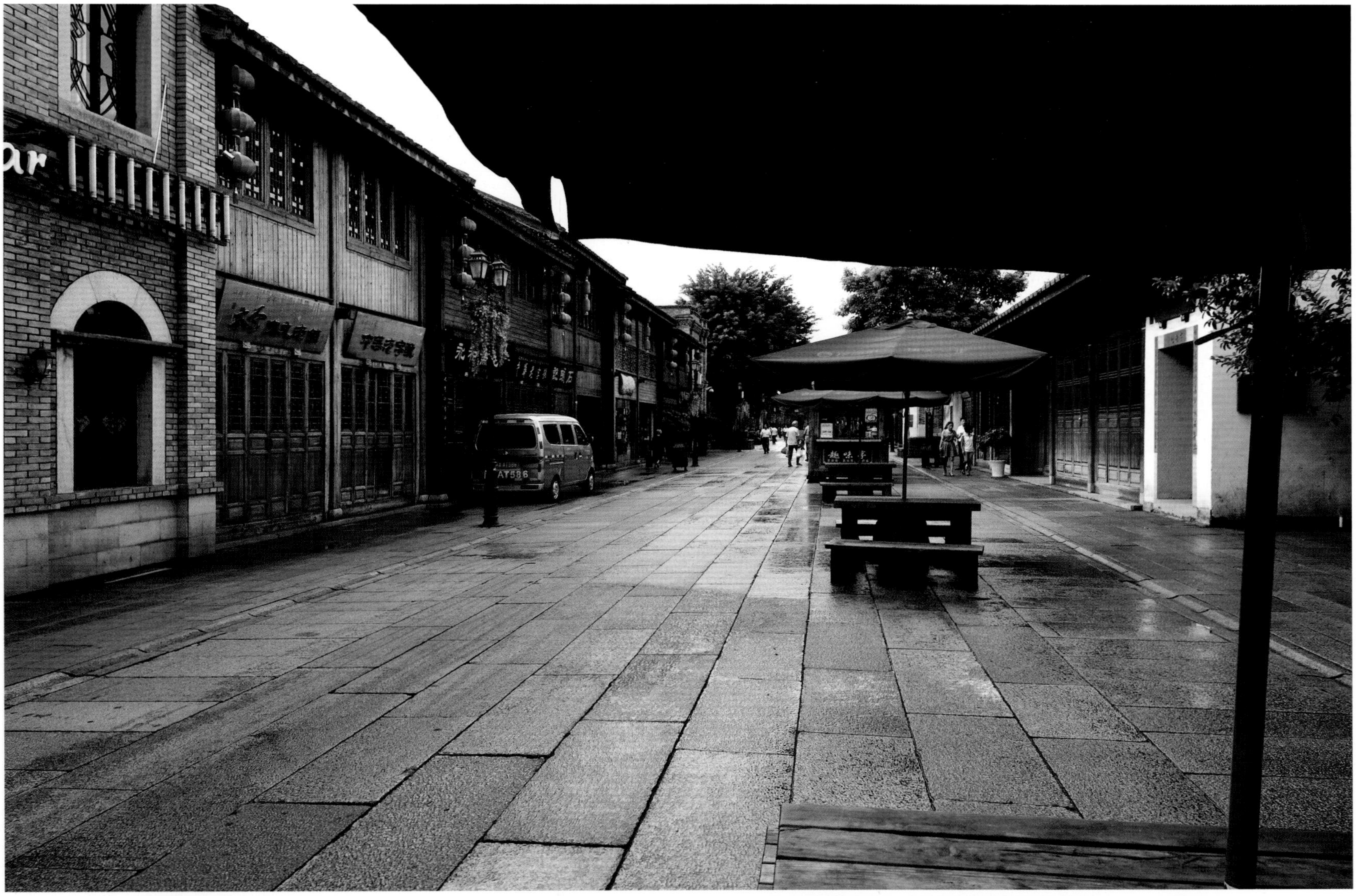

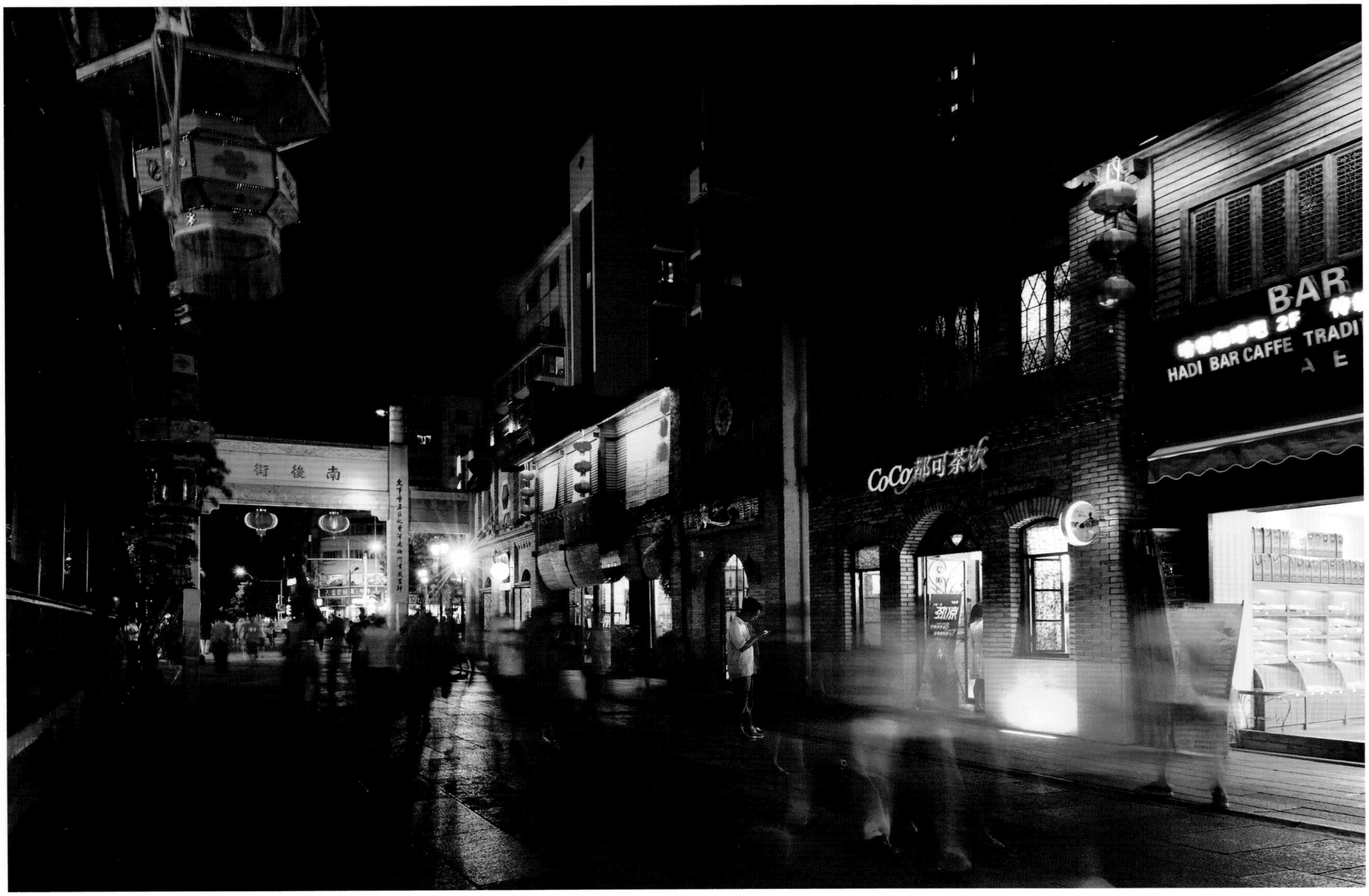
街後南
BAR
HADI BAR CAFFE TRADI
CoCo都可茶饮

Major Commercial Activities 主要商业业态

Nanhou Street has featured craft shops, history folk culture museum, experience DIY craft shops, tea art houses and nighttime tourism projects.
Antai River has featured cultural bars, open-air teahouses and featured food of Fujian.
Qixiang has museums, private studies, celebrity forums and so on.

南后街有特色工艺品店、历史民俗文化博物馆、体验式DIY工艺品店、茶艺馆、夜间景观旅游项目等。

安泰河有特色文化酒吧、露天茶座、福建特色美食等。

七巷是博物馆、私家书斋和名人讲坛等，细分为三坊七巷历史名人博物馆、福州工艺美术博物馆、福州当代漆艺馆、福州民俗展示与演艺中心、闽籍艺术家博物馆及画廊、船政文化博物馆、茶文化艺术馆、工艺品研究中心。

Operation Measure 运营措施

To develop tourism route and create tourist souvenirs, tickets, shows, accommodations, interactive experiences and so on, we remold the traditional Fujian cultural scene into movie and television shooting base.

In the creation and promotion of brand system, cultural brands, such as celebrities, Fujian customs, Fujian and Taiwan exchanged culture, are made. Meanwhile, by routine activities of museum and folk custom festivals, online special events, the brand of Sanfang Qixiang is entirely broadcasted.

开发旅游线路，推广名牌旅游纪念品、门票、演出、餐饮住宿、互动体验、区内人力车等，将富有闽都文化特色的三坊七巷传统风情场景，改建成东南沿海影视拍摄基地。

在品牌体系塑造与推广上，打造名人、闽都民俗、宗族世家民俗、闽台文化交流等文化品牌。同时，通过开展博物馆展出活动、民俗文化节等常规活动，开办网上坊巷、认祖归宗、坊巷寻宝等专题活动，举行纪念五四运动、林则徐禁烟等针对性活动，全面打响三坊七巷品牌。

Food & Beverage 餐饮类

Mujin Meatballs

It is made of taros, potatoes, pork, red dates, sesame and brown sugar and is processed into small balls of steamed meat. The meatballs look glittering and translucent with fragrance and good taste. In 2001, Fuzhou Mujin Food Co., Ltd. was founded. Semi-mechanization is adopted to volume-produce and package production.

木金肉丸

1910年姚木金采用芋头、薯粉、猪肉、红枣、芝麻、红糖等为原料，加工成一粒粒蒸熟的肉丸。肉丸晶莹透亮，如琥珀，似玛瑙，气味芳香，味道清纯，脆韧耐嚼，具有独特风味。1990年，木金肉丸曾两次漂洋过海：一次是台湾同胞购买200块，当天用飞机运去；另一次是日本侨胞购买100块，肉丸蘸上菇粉、用油炸后送到日本。2001年，姚祖健成立了“福州木金食品有限公司”，采用半机械化大批量生产木金肉丸系列食品，并精装上市。

Starbucks

Starbucks founded in 1971 is the biggest coffee chain store in the world with its headquarters located in Seattle. It sells more than 30 types of coffee, pastries, coffee machine and coffee cup. The positioning of its products is high-quality and elegance. With fresh style and fashionable leisure food and drinks, it is popular in the world.

星巴克

星巴克于1971年成立，为全球最大的咖啡连锁店，总部坐落在美国西雅图。零售产品包括三十多款全球顶级咖啡豆、手工制作浓缩咖啡、多款咖啡冷热饮料、各式糕点食品以及丰富多样的咖啡机、咖啡杯等商品。产品定位以优质、高雅餐饮为主，风格清新、味觉时尚的休闲餐饮品深受大众喜爱。

Guoshi Garden

This building has characteristics of fire and humidity protection, warm in winter and cool in summer, sound insulation and beautiful environment. It fully shows the gorgeousness of Fuzhou traditional architectural aesthetic.

国师苑

国师苑具有防火避湿、冬暖夏凉、隔音效果好、环境优美等特点，充分展现出福州传统建筑美学的辉煌，目前进驻的商铺有满记甜品等名店。

Leisure 休闲类

Chanyi Club

It is filled with slight fragrance, figures of Buddha, woodcarvings and bodiless lacquers of various sizes in the showcase. There is a small yard with boxes and seats for tasting tea, reading and chatting.

禅怡会所

在禅怡会所，淡淡的薰香弥漫其间，大小各异的佛像、木雕、脱胎漆器静静地躺在橱窗内。会所内部还隐藏着一座小院落，里面有茶艺包厢、饮品雅座，可以随意进店品茶、读书、聊天，修身养性。

Watsons

Watsons Group (Hong Kong) was founded in 1828. It has distributed its business in 34 regions with more than 8400 retail stores. Its products include health care products, beauty products, food, beverages, electronic products and so on.

屈臣氏

屈臣氏是和记黄埔有限公司旗下屈臣氏集团的保健美容品牌。屈臣氏集团（香港）有限公司创建于1828年，业务遍布34个地区，共经营超过8 400间零售商店，聘用98 000名员工，集团涉及的商品包括保健产品、美容产品、食品、饮品、电子产品、洋酒及机场零售业务等。

Qinghong

Qinghong wine has a 90-years history. At present, Hongsheng Wine Company is the biggest rice wine base in Fujian Province which has two famous brands "Qinghong" and "Minjiang".

青红

青红酒始于1920年，至今已有90余年的酿造历史。目前，宏盛酒业是福建省最大的黄酒基地，拥有“青红”和“闽江”两大强势品牌，是福建省唯一导入ISO9002质量体系认证的黄酒生产企业，投巨资打造中国红曲黄酒博物馆，为当代人留存中华千年黄酒历史，用心打造中国红曲黄酒第一坊。

Health Care 医疗类

Ruilaichun Hall

It is a large scale Chinese traditional medical comprehensive store assembling Chinese traditional diagnosis and treat, physiotherapy, health education and medicine sell.

瑞来春堂

瑞来春堂是集中医诊疗、理疗康复、健康教育、中药饮片、药品销售为一体的大型中医药综合场馆。

Dong Zhiyi's Former Residence 董执谊故居

This building is also called "Zhenji House". It is located at No. 162, Nanhou Street West. There is a remolded cut block sculpture which agrees with the "Weiyun Lu" founded by Dong Zhiyi.

董执谊别称“贞吉居”，位于福州南后街西侧162号。现新做了6扇大门，门前重新塑造了刻版印书雕塑，这与董执谊曾经开办“味芸庐”刻书坊相符。

Ermei Study (Fujian Folk Custom Museum) 二梅书屋

It is over 2,000 square meters with five parts. The fourth part was the owner's study. It was named "Ermei Study", for the two plum trees in front of it. It is a typical representative of Fuzhou dwellings in Ming and Qing dynasties.

二梅书屋占地2 000多平方米，院共五进，后门通塔巷。四进花厅原为主人书房，房前植有两株梅花，取斋名为“二梅书屋”，是福州明清时期典型的民居代表。

The Ge's Mansion (Huang Pu's former Residence) 葛家大院（黄璞故居）

With five rooms wide and three parts in depth, this 1,000 square meters building was constructed in Tang Dynasty. It was the residence of Huang Pu, a celebrity in Tang Dynasty.

葛家大院面阔五间，三进，抬梁结构，占地面积1 000平方米。始建于唐代，历代均有重修。原为唐代名儒黄璞住宅，后建为黄璞祠。1993年6月，其被列为莆田市文物保护单位。

Sanfang Qixiang Art Gallery 三坊七巷美术馆

It is a professional gallery featuring calligraphies and paintings. There are calligraphy works by Bingxin, couplet by Lin Zexu and rare works by Chen Zifen.

三坊七巷美术馆是以字画为主的专业美术馆，不仅有冰心的书法作品、“虎门销烟第一人”林则徐的书法对联，还有书画大家陈子奋的“花鸟八屏条”。

Small Yellow Building 小黄楼

Its layout is compact and its pavilions are of distinctive local features. Its environment is quiet and beautiful with a small bridge and a stream of Jiangnan garden feature.

小黄楼建筑分布紧凑，亭台楼阁具有鲜明的福州地方特色。园内小桥流水，清幽雅静，颇能体现出江南园林的特色，是福州市目前保存最秀美，且小巧玲珑的古式花厅园林。

The Yan's Parlour (Zhennan Culture and Art Museum) 鄢家花厅（桢楠文化艺术博物馆）

As a place for ancestral hall and discussing family affairs for two centuries, this building witnesses the fate of the family and the country.

鄢家花厅由福建鄢氏族人兴建并居住约两个世纪之久。作为家族的宗祠和议事之所，这座建筑曾经见证了家族和国家的运道。

The Ye's Reaidence Fujian Province Intangible Cultural Heritage Museum 叶氏民居福建省非物质文化遗产博览苑

It was built in Ming Dynasty and was repaired for several times during Qing Dynasty and the Republic of China. It is composed of a building at the east and two yards at the west with a simple plane layout.

叶氏民居始建于明朝，清朝至民国时期屡次修葺，大门朝东，外墙为民国时改建的青砖墙体。整体建筑由东面正座及西面两个侧院组成。面阔五间，为“明三暗五”格局，大厅减柱，空间宽敞；第二进三面环廊，正房沿天井北面一字排开，平面布局简洁朴实。

Zongtaozhai Art Museum 宗陶斋艺术馆

Zongtaozhai Art Museum is the largest professional exhibition in Sanfang Qixiang with completed support facilities. It is 1,500 m^2 with more than ten showcases which can exhibit over 200 calligraphies and paintings.

宗陶斋艺术馆是福州市三坊七巷最大的专业展览场馆，面积为1 500多平方米，展柜10余个，可展出字画200多幅。另有多功能厅、贵宾室、拍卖厅、仓库和安保系统等配套设施。

Taiping Street, Changsha
长沙太平街

街区背景与定位
Street Background & Market Positioning

History 历史承袭

With a very long history, Taiping Street is an epitome of old Changsha. Since Changsha became city in Warring States period, Taiping Street had been the core area of the old city for over two thousand years. The famous Qu Yuan and Jia Yi in the history used to live here. After Changsha was authorized to establish its own government office, Taiping Street was located at the south Changsha. According to the records, the space layout of Taiping Street, including Majia Lane, Hujia Lane and Xipailou, was finalized in 1818. But it was seriously damaged in "Wenxi Fire" in 1938. After reconstruction and repair, the original appearance of Taiping Street was recovered.

太平街历史悠久，是古老长沙的一个缩影。自战国时期长沙为城池以来，太平街就是古城的核心地带，历经2 000多年而未改变。在战国、秦汉时期，太平街一直是长沙城的濯锦坊，历史上鼎鼎大名的屈原、贾谊曾居住于此。到明代长沙府设立府治以后，街区一直位于长沙城南的善化县西城门小西门和大西门城内一侧。在明崇祯时期的《长沙府治》中，记载了太平街中金线巷的名字。根据清嘉庆二十二年（1818年）的《善化县志》记载可以知道，太平街街区的空间格局已经定型，太平街、马家巷、胡家巷（孚嘉巷）、西牌楼等街巷也都成型。1938年"文夕大火"之后，太平街千疮百孔、破败不堪，后经过重建、修复工作，太平街恢复原貌。

Location 区位特征

Taiping Street is located at the south of Changsha old area. Its traffic is very convenient. The whole street area with Taiping Street as principal line, extends from Wuyi Avenue at the north to Jiefang Road at the south, and is adjacent to Weiguo Street at the west and next to Sanxing Street at the east.

Changsha is one of the 24 famous historic and cultural cities by the State Council. Though it has gone through many wars and was expanded, its core area is still there.

太平街坐落于长沙市老城区南部，交通十分便利，街区以太平街为主线，北至五一大道，南及解放路，西连卫国街，东到三兴街、三泰街。街区全长375米，宽近7米，占地面积约12.75公顷。

长沙是楚汉名郡、革命圣地，国务院首批的24个历史文化名城之一，虽屡遭战乱，城区扩大，但核心城址未动，位于城市中心地带的太平街历史街区是保留原有街巷格局最完整的街区。

Market Positioning 市场定位

Taiping Street takes residence, tourism, business and culture as its main functions and focally shows the traditional culture and appearance of Changsha in the late Qing Dynasty and early Republic of China.

太平街是以生活居住、旅游服务、商业服务、文化经营为主要职能，集中体现长沙清末民初时期传统人文风貌的历史文化街区。长沙市政府力图将其打造成为一个展示湖湘文化魅力、体现千年老长沙味道的历史文化街区，使其成为长沙市的新坐标、新名片。

Street Planning 街区规划

In 2005, the government of Changsha started a protection planning project targeting at Taiping Street. The project was processed by phases according to the principal of "comprehensive planning, implementing by stages and giving priority to protection", and was planned to be completed in 2012.

The planning of 2005 determined the aim of protecting and repairing Taiping Street is to build a street focally presents the traditional culture and appearance of Changsha in the late Qing Dynasty and early Republic of China. The streets' and lanes' layout of Ming and Qing dynasties is protected to fully show the three cultural characteristics of monument of Qu Yuan and Jia Yi, heritage of business customs and headstream of the Revolution of 1911. The layout, width, style and environment of the core protection zone are protected mainly by repairing and renovation; buildings in construction controlled zone should be agreed with core protection zone; and the constructions of environment coordination zone should have more greenbelt.

2005年长沙市政府启动针对太平街历史文化街区的保护规划工程，在规划和实施过程中遵循"全面规划、分期实施、保护先行、旅游跟进"的原则分阶段进行，一期保护整治工程已竣工，二期工程于2009年完成，2012年全面完成历史街区的整治工作。

在2005年的规划中确定了太平街保护整治的目标是建设以生活居住、旅游服务、商业服务、文化经营为主要职能，集中体现长沙清末民初时期传统人文风貌的历史文化街区。保护历史文化街区明清的街巷格局，清末民初民居风貌，长沙市井的生活气息，并充分体现屈贾文化纪念地、商业民俗传承地、辛亥革命发源地三大文化特征。要求保持历史文化街区核心保护区原有的街巷格局、走向、宽度及空间尺度、建筑风貌和街巷环境，建设活动以维修、整治、修复及内部更新为主，控制建筑为二层。建设控制地带的建设活动为保持建筑风貌与核心保护区协调一致，控制建筑为四层。环境协调区的建设或更新为体量小、色调淡雅的建筑，多留绿化带。

Street Design Features 街区设计特色

Taiping Street is an epitome of old Changsha which is burdening the profound historic culture and deep folk custom of Huxiang. Walking on the street, visitors can directly feel the visual impact by the ancient buildings, also can experience breath and charm emitted by historical accumulation. In the Taiping Street, residences and shops are in traditional style and layout.

Taiping Street is a name card of Changsha and also one of the four stone roads preserved in Changsha. It is a historic street showing charm of Huxiang culture. There are not a few cultural relics and historic sites preserved at Taiping Street such as Jia Yi's Former Residence, Former Site of Gongjin Party of the Revolution of 1911, the Former Residence of Li Fuchun, Lisheng Salt Shop, Dongtingchun Teahouse and so on. The street is called home town of Qu and Jia, headstream of Huxiang culture, cradle of modern revolution and place of origin of Hunan Revolution of 1911.

太平街是“古城长沙”的一个缩影，承载着湖湘厚重的历史文化，有着浓厚的民俗风情。行走在古街，除了能直观感受到石牌坊、麻石路、封火墙、古戏台这些标志性古建筑所带来的古典视觉冲击之外，更多的是领略到一种历史积淀所散发的文气与韵味。在太平街街区内，小青瓦、坡屋顶、白瓦脊、封火墙、木门窗是这一带民居和店铺的共同特色；老式公馆保留着较为原始的石库门、青砖墙、天井四合院、回楼护栏等传统格局。

太平街是长沙的一张名片，也是目前长沙保留的四条石板街之一，是展示湖湘文化魅力、体现千年老长沙韵味的历史文化街区。太平街街区保存有较多的文物历史古迹和近代历史遗迹、众多的行栈、旅店和戏园、茶楼。如贾谊故居、长怀井、金线街的麻石街、西牌楼旧址、辛亥革命共进会旧址、四正社旧址、李富春故居、乾益升粮栈、利生盐号、洞庭春茶馆、宜春园等，其被称为屈贾之乡、湖湘文化的发源地、中国近代革命的摇篮、湖南辛亥革命的策源地。

楚天會所
湘菜 棋牌
楚天藝术會所
Friend's

百适甜品

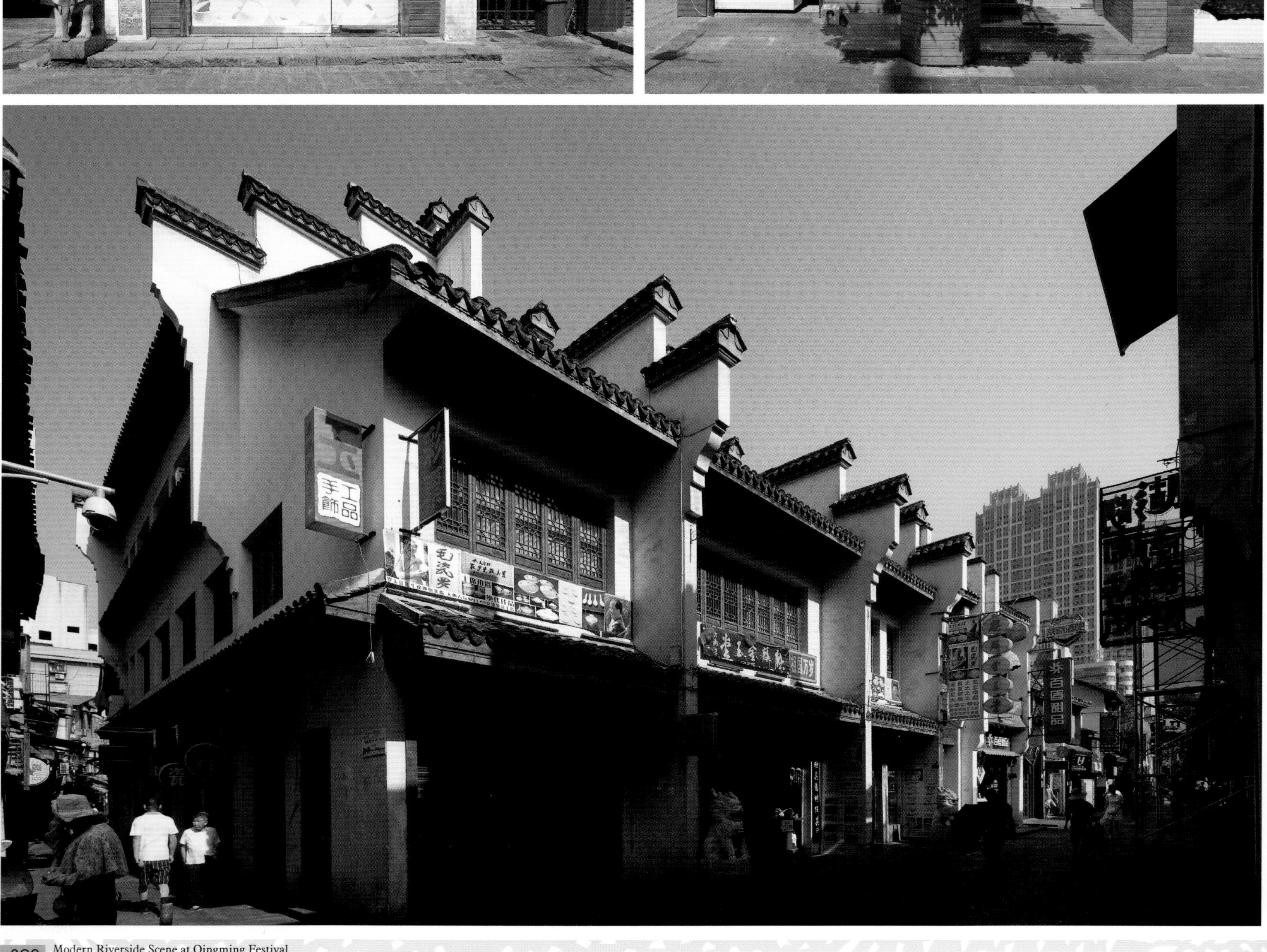

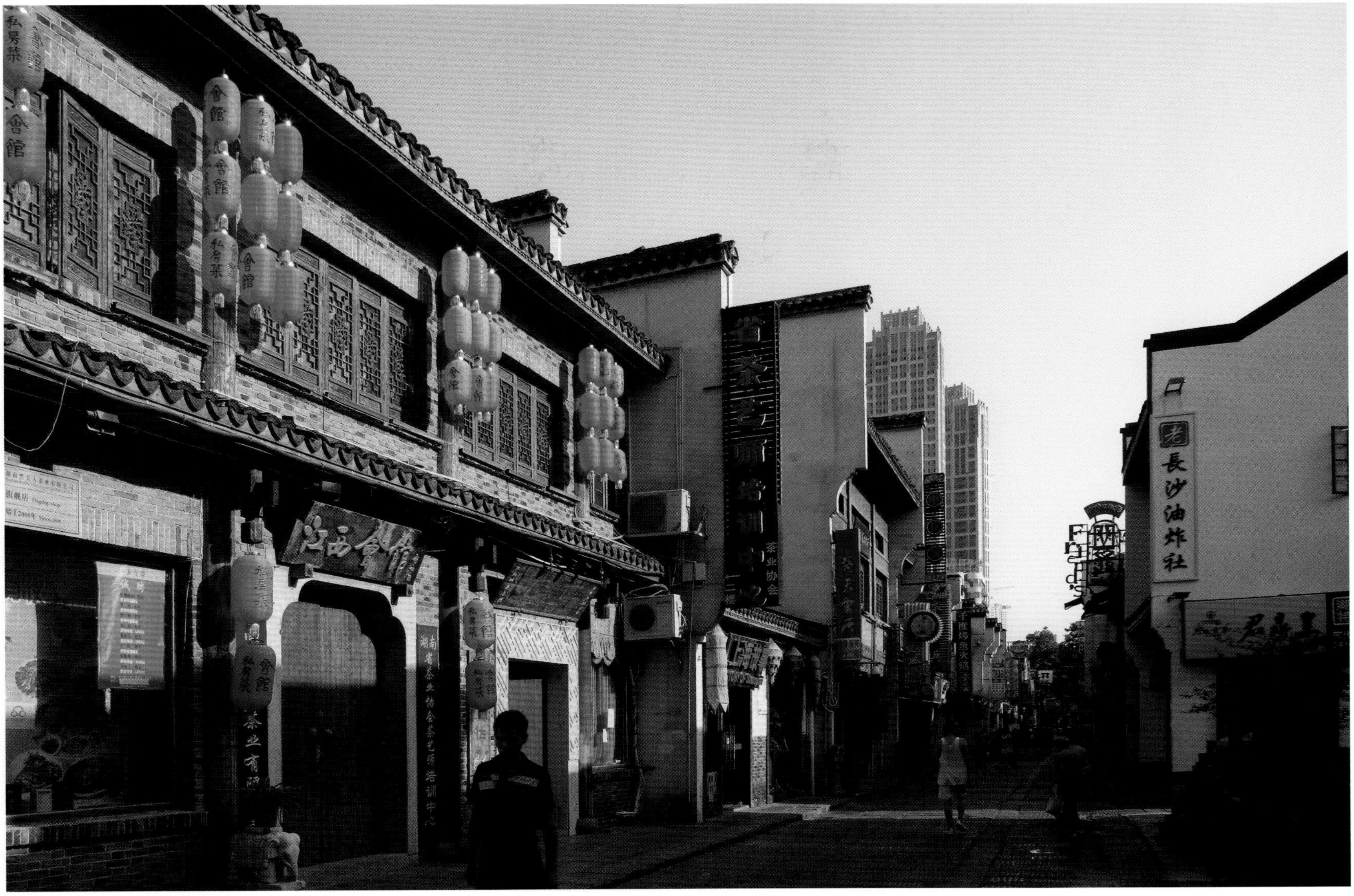
老
長沙油炸社
湖南省茶业协会茶艺师培训中心

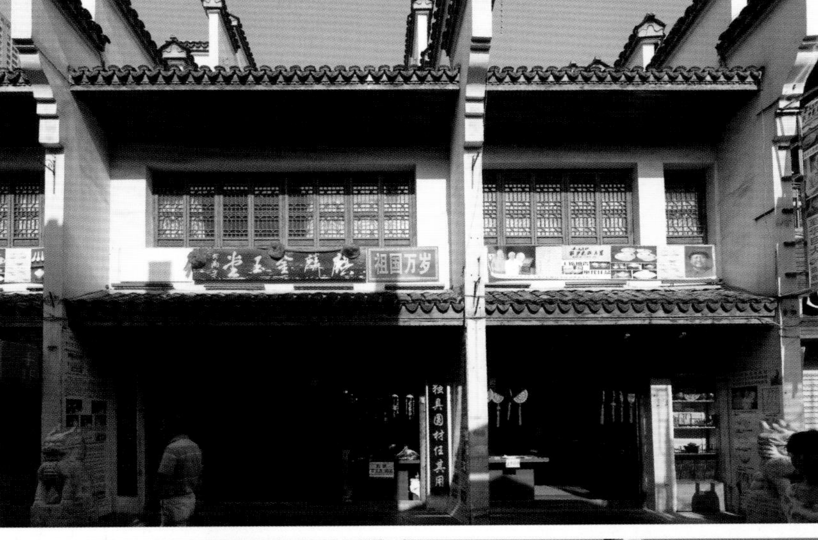

祖国万岁

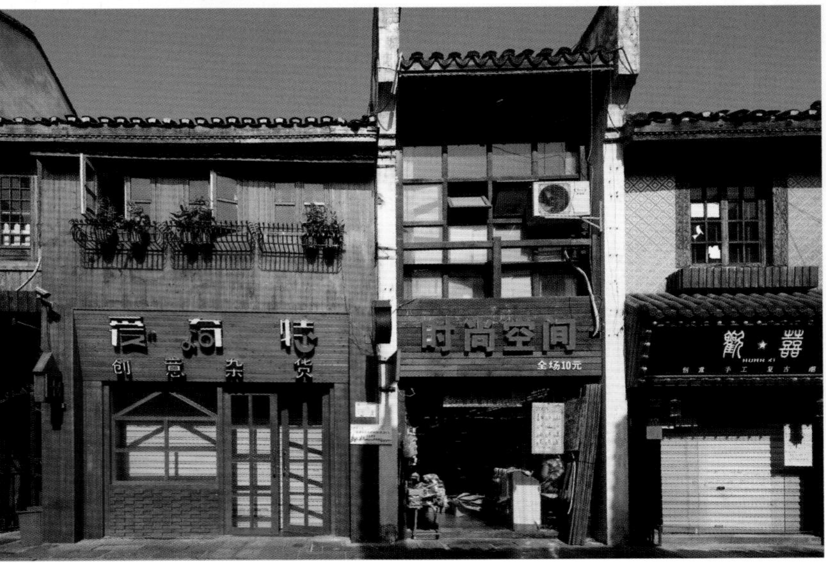

创意杂货
时尚空间
全场10元

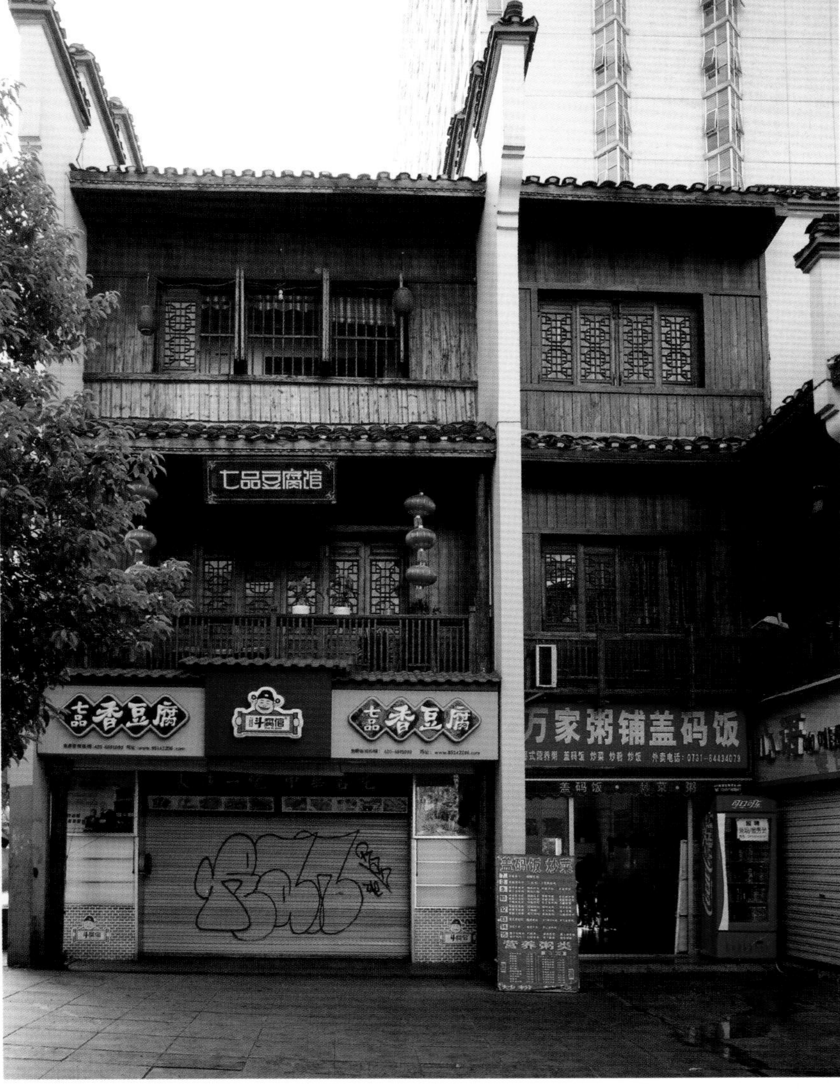

七品豆腐馆
古品香豆腐
古品香豆腐
万家粥铺盖码饭

没有吧?

冬林府

楚天藝术會所

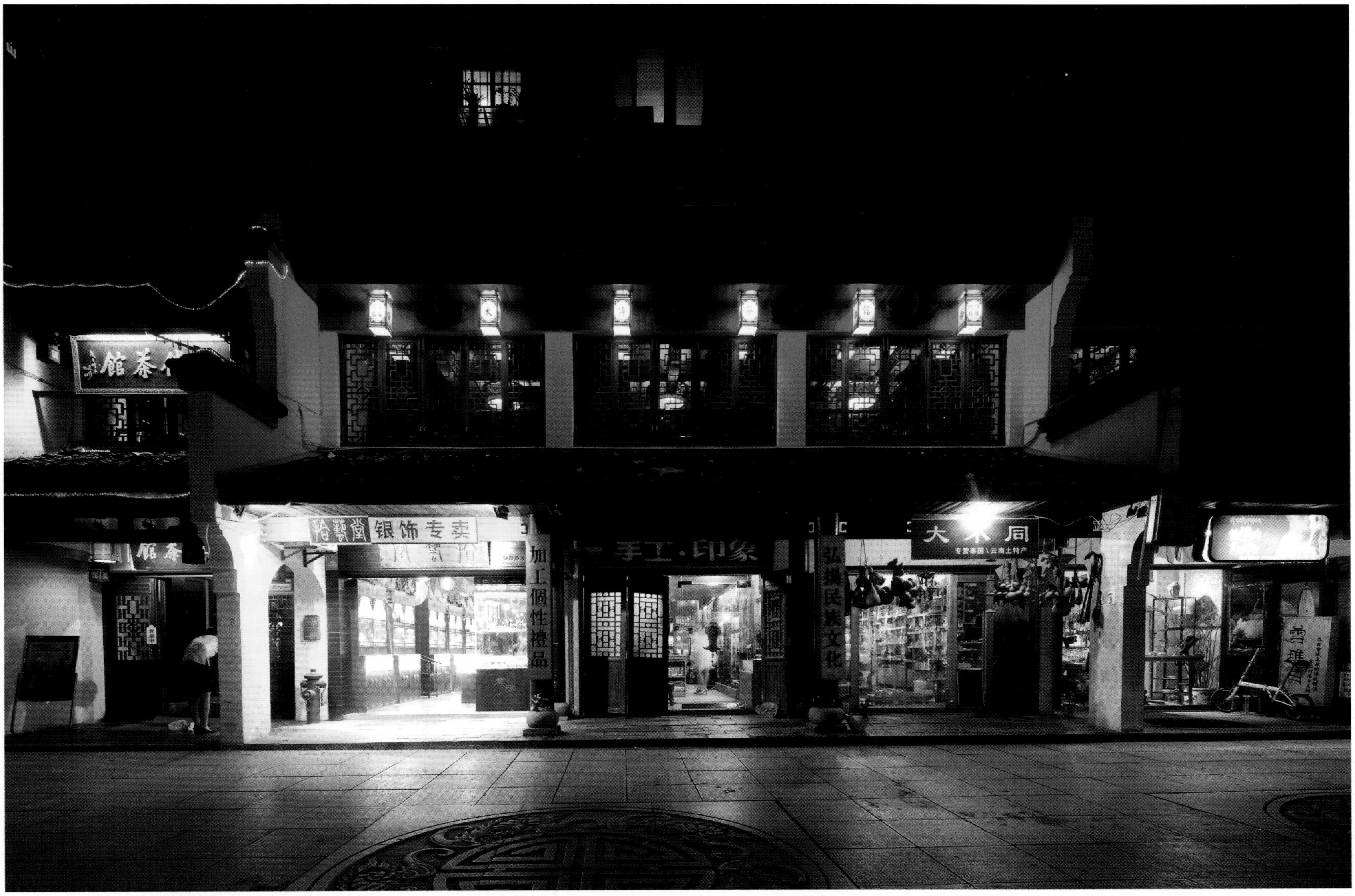
银饰专卖
加工個性禮品
手工·印象
弘揚民族文化

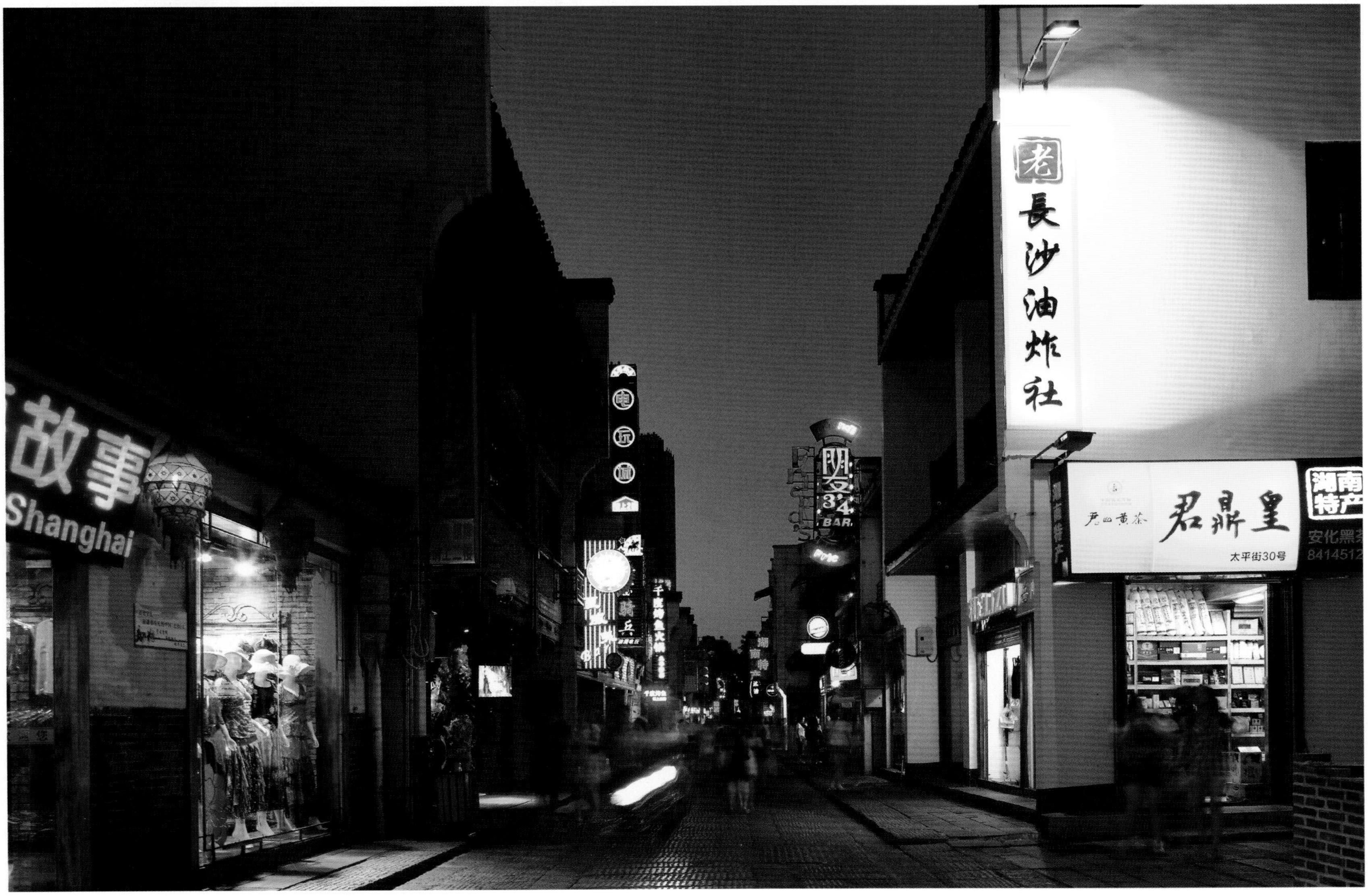
故事
Shanghai
老
長沙油炸社
君鼎里
太平街30号
湖南
特产

Major Commercial Activities 主要商业业态

The major commercial activities in Taiping Street include traditional featured business, restaurants, traditional culture industries, featured crafts and business services for historic street. Taiping Street is a traditional commercial street. Old shops like Jiuzhitang Pharmacy and Old Yang Mingyuan Eyegalasses Store, and businesses like paper-cut traditional painting are restored. Xipailou Street is a restaurant culture leisure street. Many old famous restaurants and teahouses are restored. Jinxian Street is a local culture and custom street. Businesses of folk crafts, such as embroidery, paper-cut and so on, are restored.

太平街历史文化街区的主要商业业态有传统特色商业、餐饮业、传统文化产业、特色手工艺品以及为历史街区服务的商业，街区内老字号应有尽有。太平街为传统商业街，恢复了九芝堂药店、老杨明远眼镜店、和记多绸布庄、乾升泰干货店、怡丰商场、宜春园茶馆、宜春园戏园等老字号和湘绣、剪纸、传统字画、旧书文物、手工艺品等商业。西牌楼街为传统餐饮文化休闲街，恢复了马明德、半雅亭、李合盛、德园、玉楼东等老字号餐馆，洞庭春、玉壶春、涌湘亭、听月楼等老字号茶馆，特色餐饮和特色小吃以及老字号戏园。金线街为地方文化民俗文化街，恢复了汉新和纸号、晋康纸行、彭三和笔庄、大昌祥颜料号以及经营捞刀河、沙坪、长沙县、望城的湘绣、剪纸、民俗工艺品等商业。

Featured Commercial Area 特色商贸区

Apart from time-honored shops, the Taiping Street also pays lots of attention to create a modern featured commercial area. For example, the Women World Shopping Mall.

长沙太平街中除了老字号外，还注重打造现代具有特色的商贸区，例如以女性为主题的女人世界商城。

Women World Shopping Mall

Women World shopping Mall is a feminine theme shopping mall. The Women World Shopping Mall at Taiping steet emphasizes on protecting historic relics in Changsha and its building style continues the antique charm of Taiping Street to create a tourist type of shopping. Women World shopping Mall brought in a 5A intelligent parking system to solve problems of traffic jam and inconvenience of parking.

女人世界

女人世界是深圳市女人世界连锁股份有限公司和长沙女人天地项目开发商打造的长沙首家深圳“女人世界”姊妹店，是一家女性主题商城。太平街中的“女人世界”注重保护长沙历史古迹，在建筑风格上沿续太平街的古韵，打造的是一种游乐式的购物。女人世界在解决交通拥挤、停车不便的问题上引进了5A智能化停车系统。

招商热线:84321392

新靓歌坊KTV
女人世界长沙购物中心
招商热线 84321392 84321390
湖南特产

世界长沙
人世界长沙
女人世界长沙
世界长沙购物中

Food & Beverage 餐饮类

Starbucks

The Starbucks in Taiping Street is designed elaborately. Standing on the street, visitors can see the Qing style wooden frame windows. And as they go into the shop, they can see brown and wooden colored furnishings. Chinese style is integrated into every detail of this space, for example, sofa armrests are wrapped with copper-colored wire nails, dark coffee chairs are decorated with ancient coin patterns, and the hall is filled with Chinese classic music…

星巴克

太平街中的星巴克是经过精心设计的，站在街边可以看到它清代风格的木格门窗，咖啡馆内部是满目的棕色和原木色，中式风格融入这片空间的每一处转角和细节之中，沙发扶手上包裹着古铜色的圆钉，深咖色座椅上布满了天圆地方的古钱图案，一张明清风格的长柜作为调味吧台，复古的壁灯渲染出舒适的氛围，客厅中还流淌着中国古典音乐，与咖啡的香味一起融入客人的心田。

Agan Teahouse

It is the first pure Chinese style teahouse that mainly sells Hunan black green tea and famous high-quality tea in Hunan. Apart from selling teas, it also provides customers with communication place for art activities.

阿甘茶馆

阿甘茶馆是湖南首家以经营湖南黑茶及湖南名优茶为主的纯中式茶馆，除主营湖南名茶外，茶馆还为茶友提供进行琴棋书画活动的交流场所，是颜家龙老先生书画作品的唯一指定的交流场馆，茶馆同时也为有缘人提供佛事活动、禅茶斋席的预约。

Dabing Pungent and Spicy Meat

It is located at No.9 Taiping Street and has dozens of chain shops now. It has a good material allocation system and a good reputation. It is a pungent and spicy meat manufacturer of a long history in Changsha.

大兵麻辣肉

大兵麻辣肉位于太平街9号，现已拥有数十家连锁店。它拥有良好的配货机制，口碑极好，是长沙历史悠久的生产麻辣肉的厂家，主要经营麻辣肉、牛肉干、肉脯等特色地方小吃。

Qipin Tofu Mill

Qipin Tofu Mill is a shop specializing in Qipin Tofu. Qipin Tofu has seven favors which are unique by different workmanships. Its products are very popular in customers.

七品豆腐坊

七品豆腐坊是专营七品豆腐的商铺。七品豆腐风味独特，汇集“卤”“酱”“炸”“煎”“烤”“煮”“酿”等香豆腐制作工艺，品种有七款，深受食客喜欢。

Jiangxi Guild

Jiangxi Guild is a two-floor imitated antique building. Its first floor is for selling tea and tasting tea, and its second floor is a Hunan Tea Committee Tea Art Specialist Training Center.

江西会馆

江西会馆是一栋二楼的仿古建筑，门面与花制作并列，一楼是卖茶、品茶的场所，二楼是湖南省茶叶协会茶艺师培训中心，协会会长曹文成担任校长，教学场地380平方米左右，为茶叶行业培养合格优秀的茶艺师人才。

Taifu Teahouse

Taifu Teahouse combines modern vision with classic art. Its general design integrates a large number of new classic elements. There are various kinds of seats in the teahouse with elegant environment and graceful arrangement. The teahouse adopts steady warm colors as basis to create a warm and friendly feeling.

太傅茶馆

太傅茶馆是一家现代视觉与古典艺术相结合的茶馆，整体设计融入了大量新的古典元素。店内拥有各种类型的卡座，环境优雅，结构大气。二楼临窗的光线很好。茶馆在色彩控制上以稳重的暖色调为基础，配合局部光源处理，显得亲切温馨。

Crafts 工艺类

Fenghuang Wax Dye House

Wax-dyeing means using specially-made wax knife to paint various kinds of patterns on white cloth with melting beewax. After the painted white cloth is dyed, cleaned and dried, a colorful printed cloth is made.

Fenghuang Wax Dye House is located at No.174 Taiping Street. Its products are from Fenghuang and three fourth of them are wax-dyeing products.

凤凰蜡染坊

蜡染是用特制的蜡刀，蘸上熔化的蜂蜡在白布上画出各种花纹图案，然后将画好的白布染色，再将染了色的布经沸水去蜡、清水漂洗后摊平晾，而制作出的多姿多彩的蜡染花布。蜡染大多以蓝靛液浸染，呈蓝白相间的效果。由于点、线、面的配合有致，宾主、大小、蓝白的疏密得当，自然生成冰纹的虚实变化，使白底蓝花或蓝底白花显得更加清秀淡雅，富有韵味。

凤凰蜡染坊在太平街174号，其产品均从凤凰运来，商铺中四分之三是蜡染产品。

Naxi Baozang

Naxi Baozang is located diagonally across Jia Yi's former residence. It is a private enterprise specializing in market running of tourism development. It entered Changsha market in 2012 mainly selling crafts, clothes and jewelry.

纳西宝藏

纳西宝藏位于贾谊故居斜对面，是一家专门从事旅游开发市场经营的民营企业，2012年进入长沙市场，主营工艺品、服装、珠宝等产品。

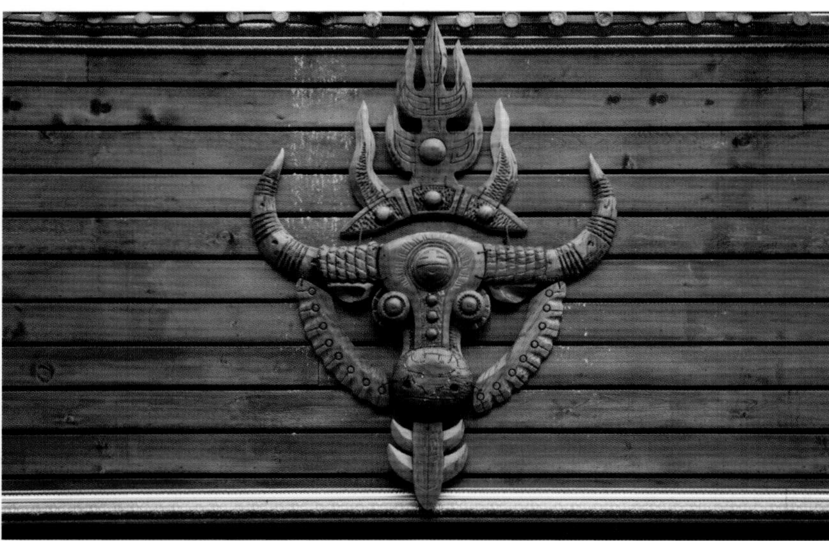

Rongbaozhai

Rongbaozhai with a three hundred years history is a time-honored shop known at home and abroad. Its main products are modern painting and calligraphy, calligraphy supplies, folk crafts and so on.

荣宝斋

荣宝斋是驰名中外的老字号，迄今已有三百年历史。荣宝斋（长沙分店）主要经营艺术品交易拍卖、近现代书画原作、木板水印，各种画谱、画册、画片，揭裱古今书画、文房用品、民间工艺品等。

Traditional Chinese Medicine 中医药类

Yaowangtang

Yaowangtang has been settled in Taiping Steet since 2008 and was the only traditional Chinese clinic which has its own traditional Chinese medicine decoction pieces production base in Hunan at that time. It provides over one thousand kinds of traditional Chinese medicines and has many professors taking in and treating patients.

药王堂

药王堂于2008年落户长沙太平街，是当时湖南唯一一家拥有自己中药饮片生产基地（三湘饮片厂）的中医馆。馆内有近1 000平方米的营业面积，有千余种中药品种，并有湖南省多位专家应诊，定期邀请著名专家学者举办中医药防治亚健康养生的知识讲座，弘扬中医中药文化。

Culture Facilities 文化设施

Jia Yi's Former Residence 贾谊故居

Jia Yi's Former Residence is located at Taifu Li, Taiping Street. It was the residence of Jia Yi in the Western Han Dynasty and is called the origin of Huxiang Culture. With about one thousand years history, it is the oldest relic in Changsha. From Chenghua Period of Ming Dynasty, its layout is of unification of ancestral hall and dwelling. During more than one thousand years, Huxiang people have rebuilt and repaired the Jia Yi's Former Residence for several hundred times.

贾谊故居位于太平街太傅里，是西汉贾谊之宅，被称为湖湘文化的源头，至今已有千年历史，是长沙最古老的古迹。贾谊故居从明朝成化元年以来就是祠宅合一的格局，宅内雕有木像一座，保存至今。现在故居上的祠匾是赵朴初先生的墨迹，祠两边的文字均是清朝时期湖南巡抚所写。在一千多年的历史中，湖湘人民对贾谊故居的重建和维修次数多达百余次。

Taiping Granary 太平粮仓

Taiping Granary was founded in Xianfeng Period of Qing Dynasty by Zhu Changlin, a wealthy merchant in Changsha. It is a Chinese- Western style modern building. Its front part is shop and barn, and its back part is the dwelling of its owner.

太平粮仓老字号叫乾益升粮栈，由清末长沙富商朱昌琳于清咸丰年间开设。它是一座中西合璧的近代建筑，其立面造型采用西洋近代建筑手法，雨山墙的做法又具有明显的长沙地方特征。前栋为铺面和粮仓，后栋为粮栈主人住宅。

Yichunyuan Stage 宜春园戏台

Yichunyuan Stage is located at the joint of Taiping Street and Xipailou Street. It combined teahouse and traditional Chinese theatre and was very popular in Changsha in Guangxu Period of Qing Dynasty. The stage is built of black semicircle-shaped tiles and red woods, and partly decorated caving patterns and gold foils. Its appearance is elegant and gorgeous.

宜春园戏台位于太平街与西牌楼街交汇处，与清末著名戏园宜春园、同春园一脉相承，集茶馆与戏园于一体，清光绪年间风靡长沙城。戏台四角的屋檐飞翘，翘檐上有精致的龙头雕饰和七个或蹲或立的压脊兽。整座戏台以青黑色筒瓦与朱红色木构相搭配，局部雕花贴金箔，沉稳大气而不失华美。

引导指示系统

Guidance & Sign System

Shangxiajiu Pedestrian Street, Guangzhou
广州上下九步行街

街区背景与定位 Street Background & Market Positioning

History 历史承袭

Shangxiajiu Pedestrian Street had become a commercial area since the 6th century. During Ming and Qing dynasties, surrounding area of Shangxiajiu Pedestrian Street was flourished. After the Thirteen Hongs was burnt up, the business activities moved to Shangxiajiu gradually. In late Qing Dynasty, Shangxiajiu was prosperous and became an important gateway for Guangzhou to trade home and abroad.

As hundreds of years pass by, Shangxiajiu Pedestrian Street together with nearby markets has composed an important business net which satisfies different merchants and citizens.

广州上下九步行街早在6世纪就已成为商业聚集区，印度高僧达摩曾在此登岸传教，因而此地得名“西来初地”。明清时期，随着接待外国使者和商贾的怀远驿的设立、大观河的开通以及十三行的兴旺，上下九步行街周边的商业兴旺发达。后十三行遭火灾烧毁，商业活动逐渐转入上、下九路。到清朝末期，上下九步行街十分繁华，周边地区衍生出许多与之相关的专业集市，成为了广州与全国及海内外进行贸易往来的一个重要窗口。

时过百年，上下九步行街商业气氛日益增旺，其周边的集市也已发展成为该区的一个重要商业网络，构成了一个近2.5平方千米的庞大商业网络，商品种类繁多，满足了不同类型的货商和市民的需求。

Location 区位特征

Shangxiajiu Pedestrian Street is located at Xiguan area of old district in Guangzhou. It starts from Shangjiu Road and Xiajiu Road at the east and extends to Dishifu West. It is 1,200 meters long with hundreds of shops. This street, which is the first commercial pedestrian street in Guangzhou ratified by National Ministry of Commerce, has a profound culture connotation that assembles Lingnan architecture culture, cooking culture and folk custom.

上下九商业步行街位于广州老城区西关一带，东起上、下九路，西至第十甫西，横穿宝华路和文昌路，全长1 200多米，1995年被国家商业部批准为广州市第一条商业步行街。街内商业繁荣，全路段商铺林立，有数百家商铺与数以千计的商户；人流量大，日客流量高达40余万人次；文化气息浓厚，荟萃了岭南文化中的岭南建筑文化、饮食文化和民俗风情。

Market Positioning 市场定位

Shangxiajiu Pedestrian Street is a reflection of Xiguan style, one of the three biggest traditional commercial centers in Guangzhou, and the first commercial pedestrian street in China. It is positioned as a modern commercial pedestrian street assembling food, shopping, entertainment and tourism.

上下九步行街是西关风情最直接的映画，是广州市三大传统繁华商业中心之一，也是中国第一条开通的商业步行街。街区中汇集了岭南特色的骑楼商市、老街名店，既是广州市的美食据点，也是外地游客寻找传统广州的第一着眼点。因而，上下九商业步行街的市场定位是集饮食、购物、娱乐、珠宝首饰、旅游于一体的现代商业步行街。

Street Planning 街区规划

The traditional Xiguan arcade buildings along the street were built in Qing Dynasty. They blend in south European architecture style and their decorations take reference from north Manchu style. This old building group is practical and artistic and is still maintaining its commercial function.

There are many historical sites at Shangxiajiu Pedestrian Street which pervade culture breath in this place. Now the exhibition and sales hall of Shifu Bookstore and Ping'an Theatre add new culture oasis to the street.

After overall reconstruct, Shangxiajiu Pedestrian Street has become more shining. Though on both sides of the street are still traditional arcade buildings, stunning lightings are added.

There are also sculptures that reveal the traditional life of Xiguan. Besides, Lichi Bay Culture Exchange Association even employed a person to play staff selling Jigonglan, which is a featured snack in Guangzhou, to keep Jigonglan culture.

上下九步行街历史悠久，街内具有传统西关房屋建筑风格的骑楼连绵千里，从第十甫一直延伸至上九路，这些建筑始建于清代，融合了南欧建筑的风格，在装饰上借鉴了北方满族装饰风格。这一古老的建筑群现在仍在维持着商业功能，使商户、顾客在任何天气下都可以进行商业活动，既实用又美观，吸引了海内外无数游客观光旅游，日客流量数以十万计。

上下九的大街小巷上留有许多历史遗迹，如昔日十三行富商在下九路组建的“文澜书院”，清同治年间状元梁耀曙曾居住在湛露直街，曾以“西关古坛”“霜花小苑书画展” 驰名的陶陶居，群众自娱自乐的粤曲演唱“私伙局”等，文化气息弥漫于街头巷口之间。现在，十甫书店的荔湾雅苑书画展销厅和上演粤剧的平安戏院成为了这条步行街上新的文化绿洲。

经过全面修建后，上下九步行街焕发新颜。马路两旁仍然是传统的骑楼式建筑，大街小巷中依然可以见到西关大屋和麻石街巷。马路上新建起300多米长的大型射灯喷画《羊城景廊》和2 000多个灯笼和彩灯，一到晚上，在霓虹灯光的映衬下，上下九步行街平添几分姿采。

街区中还设置有一组组反映西关传统生活的雕塑，如“凉茶档”“人桥”“老车夫”“门前倩影”等，这些代表广州文化的雕塑唤醒了市民的传统广州情结，西关情愫流淌心中。另外，荔枝湾文化交流协会为保存鸡公榄文化，专门请人扮演卖鸡公榄的工作人员，他们头戴尖头扁帽，两颊涂着红胭脂，身上背着五彩大公鸡，用唢呐模仿公鸡叫着“卖榄”，体现出浓浓的西关民俗风情。

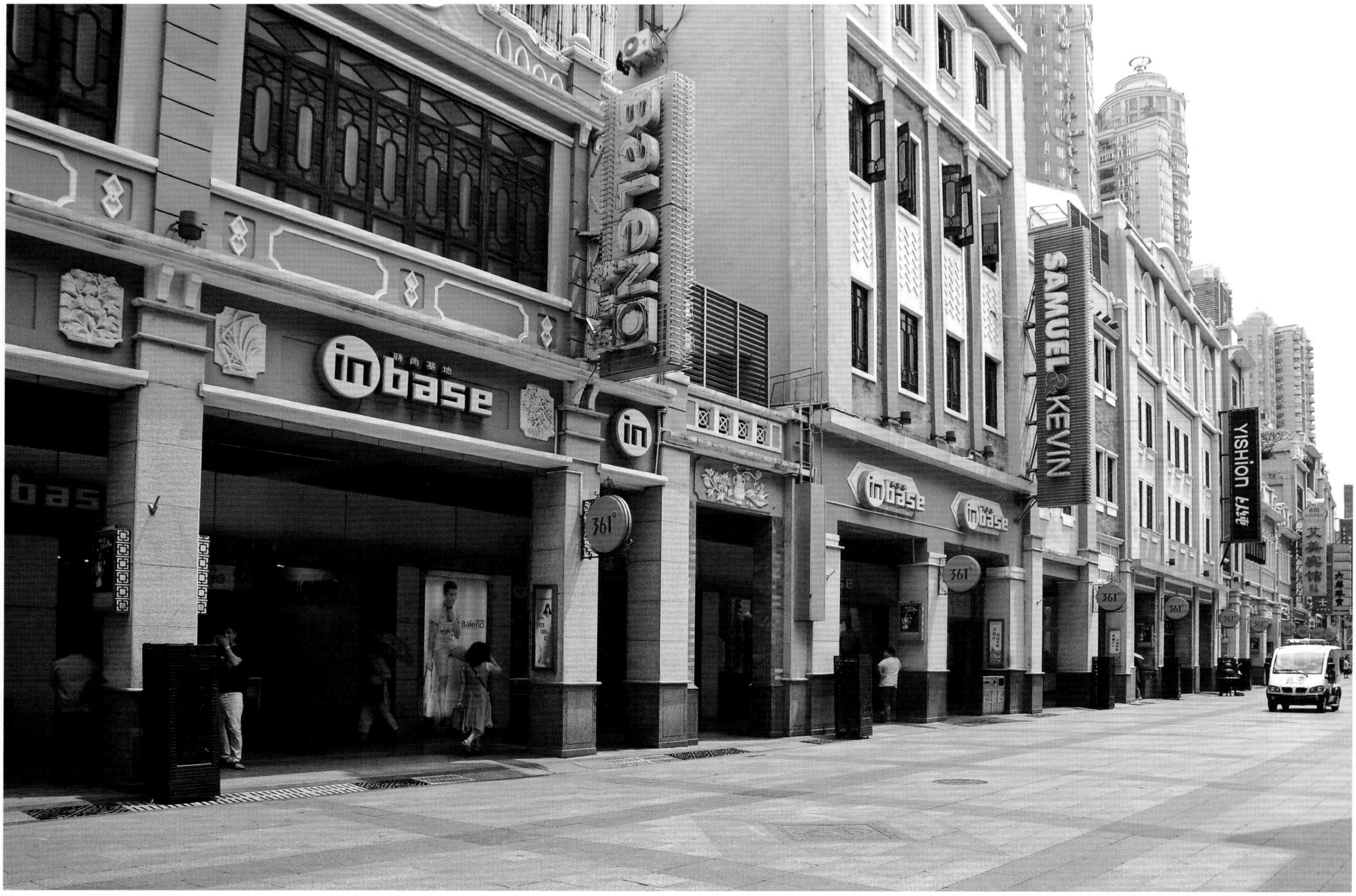
inbase
時尚基地
Baleno
361°
SAMUEL&KEVIN
YISHION

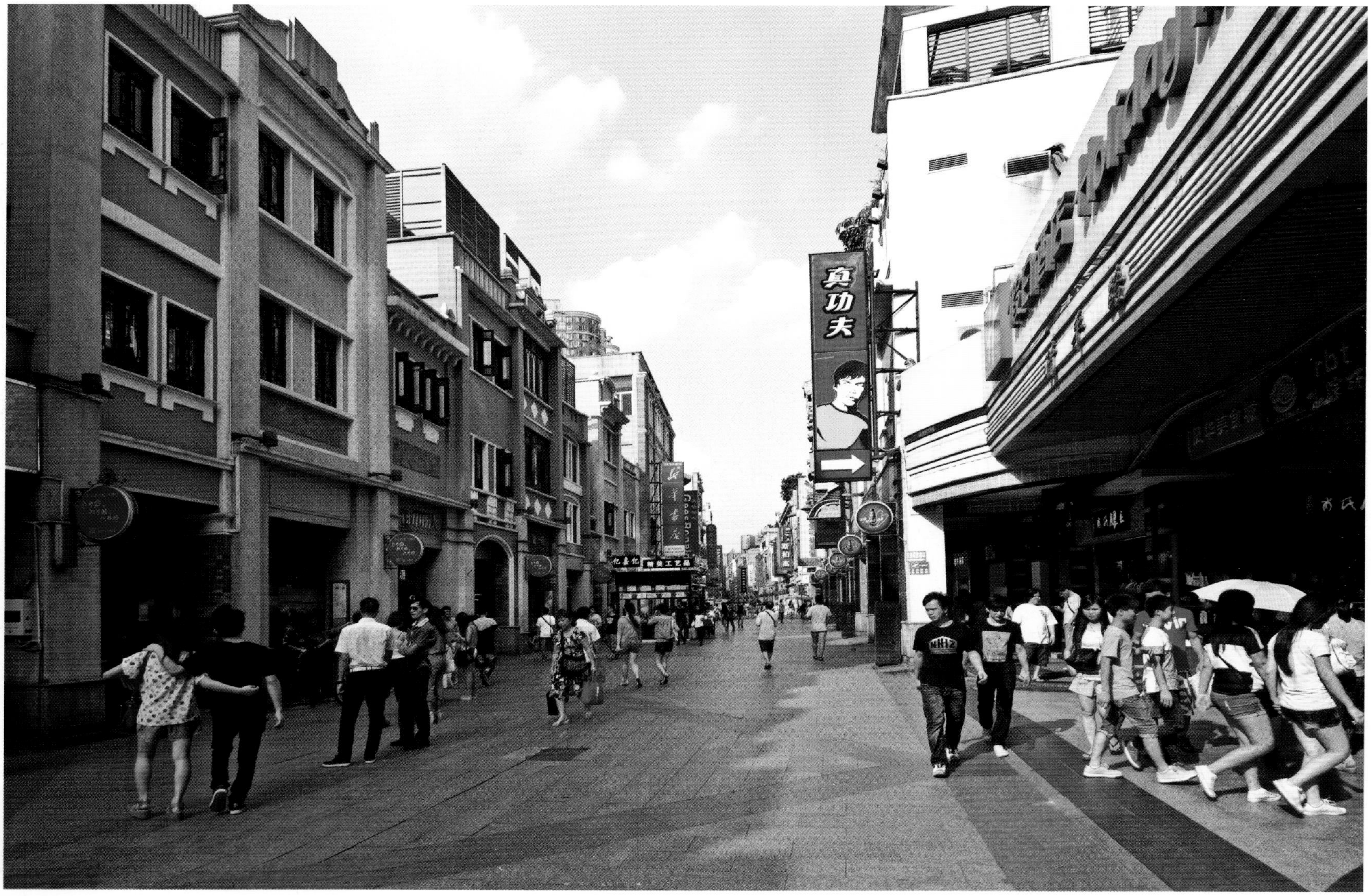
真功夫

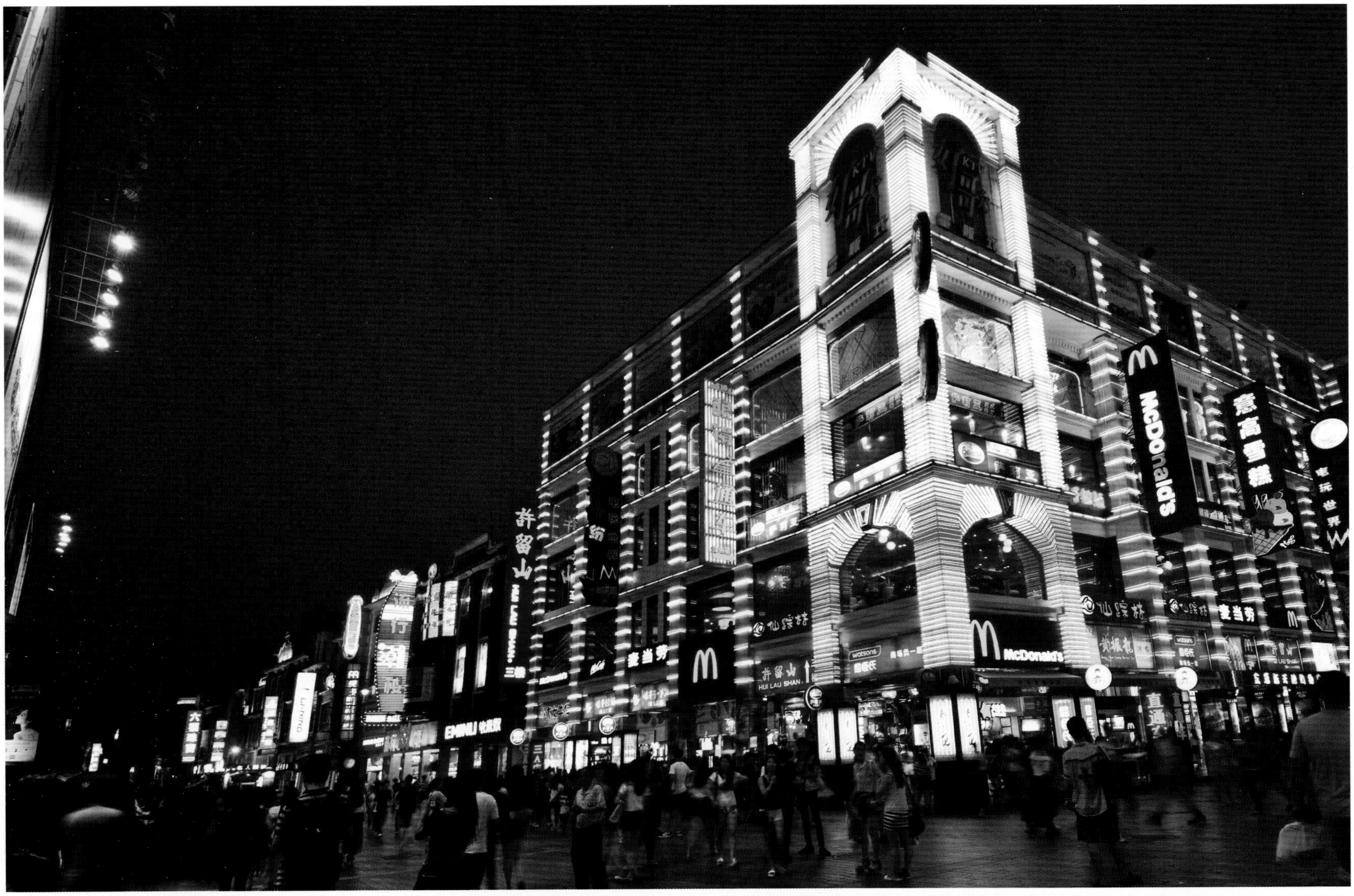
McDonald's
許留山
HUI LAU SHAN
意高雪糕
麦当劳

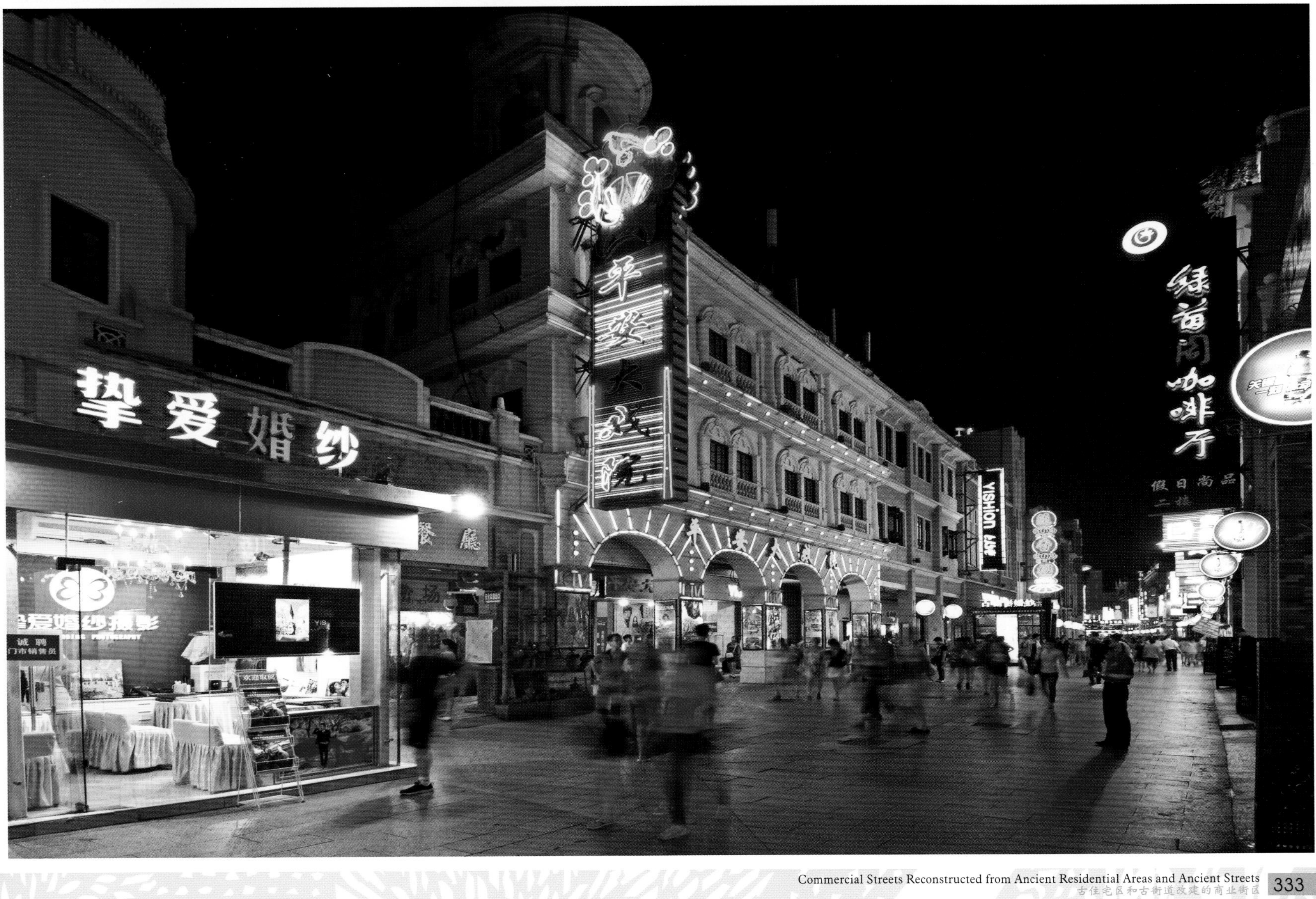
挚爱婚纱
绿茵阁咖啡厅
假日商品
二樓
餐廳

真功夫
假日酒店 Holiday Inn

必胜客
必胜客欢乐餐厅
DISHIQIAOCHE
必胜客欢乐餐厅

珠玑路
ZHUJI LU
肥佬肥婆

店
XINHUA BOOKSTORE

乐路

ZIYOUREN
休·闲·食·品·专·卖
东鹏特饮

女子医院
McDonald's
广告位招租

www.zhouheiya
关注周黑鸭官方微博
名牌运动鞋折扣店

Major Commercial Activities 主要商业业态

There are over two hundred shops in Shangxiajiu Pedestrian Street. They are mainly shops of clothes and accessories, and snack bars and electronic product shops as complement. There are old famous shops such as Yong'an Department Store, Guangzhou Textile Market, Jinhua Bedding and so on. In addition, there are tens of restaurants of different sizes such as old famous restaurant — Taotaoju Restaurant and Lianxianglou Restaurant, and national super restaurant — Guangzhou Restaurant, as well as some snack bars such as Nanxin Dessert, Oucheng Noodle and Xiguan Household.

上下九步行街内共有商业店铺200余家和数千商户，以服装饰品商铺为主，小吃店、电子产品商店为辅。上下九步行街中布满了百货公司和服饰店，有永安百货公司、广州服装店、鹤鸣鞋帽店、妇女儿童百货商店、广州织坊商场、锦华床上用品店等老字号，一些顶级品牌的新货旧款常年在“清仓大热卖”。除了这些，步行街上还有大小食肆数十家，既有百年老店陶陶居、莲香楼，又有“国家特级酒家”广州酒家，还有一批经营西关名小吃的南信甜品店、欧成记面馆、西关人家等特色小吃店，充分体现了“食在广州，味在西关”的饮食文化风情。

Featured Commercial Area 特色商贸区

Shangxiajiu Square

Shangxiajiu Square locates at the joint of Shangxiajiu Road and Kangwang Road with Liwan Plaza, Dongji Plaza, Saibo Plaza and Minghui Jewelry Jade Plaza at its four corners. Promotion activities are often held in this square. The design of the square respects the traditional space characters of old district and satisfies the demand of city function, therefore, it expands Xiguan culture and at the same time improves the leisure function of the pedestrian street.

Operation Measure

In operation and promotion, Shangxiajiu Pedestrian Street takes dominant strategies as follows: attracting merchants and shopkeepers to fill the street and taking actions to flourish and promote the street.

Strategies of attracting merchants and shopkeepers

1) Three sales points–large visitors flow rate, consuming project of youths, low threshold and risk.
2) Support by creative plan. Developer and operator have the strong ability to design plans and strong executive force.
3) Breakthrough at key points. Key points such as fashion brands, famous snack favored by youths and so on.
4) Mobilizing individual merchants. By advertising and widespread canvass to attract individual merchants.

Strategies of flourishing and promoting the street:

1) Target at destination consumers and pertinently choose channel of spread and propaganda carrier.
2)Continuously plan various kinds of theme activities that agree with targeted consumers, such as shows, participatory activities, and sales promotion which are purposeful, initiative.
3)Attract customers by connotation. Make sure there are excitements for targeted consumers at their first visit.
4)Random consumption. At the early stage, effort should be made to attract customers to spread the brands and cultivate a group of loyal consumers.
5) Pay attention to both customer source of Shangxiajiu and area targeted customer source. Attract more customers by popularity and public praise.
6)Snowball effect. Leave targeted customers a profound impression and then make use of their consumption habit of shopping in groups to bring more business.

上下九广场

上下九广场位于上下九路和康王路的交界处，广场的四个角落分别是荔湾广场、东急广场、赛博广场、名汇珠宝广场，还零星有一些雕塑，广场上经常有产品推广活动，人流量特别大。广场设计尊重原有旧城区的传统街道空间特性，满足新的城市功能需要，起到弘扬西关文化、建设民俗风情、提高步行街休闲功能的作用。广场还有一批经营西关名小吃的南信甜品店、欧成记面馆、西关人家等特色小吃店，充分体现了“食在广州，味在西关”的饮食文化风情。

运营措施

在运营推广上，上下九步行街采取的主要策略是：①旺丁、旺场；②招商，满场。其中招商推广在前，旺场推广在后，开业后前后并举，互为推动。

招商推广策略

在招商推广策略上主要有以下四点：

1）三大卖点吸引。上下九街区人流量大，蕴含大商机；以年轻人为消费主体，商机无限；低门槛、低风险，在有保障的情况下创富。以这三大卖点吸引商家入驻。

2）创意策划支持。开发运营商具备旺场策划能力和执行力，有大手笔旺场推广实例，具备确保旺场经营的广泛的社会资源和人脉资源。

3）锁定重点突破。锁定潮流商铺的品牌生产厂商或经销代理商；锁定具备日、韩、中国港澳台概念的客商；锁定青少年偏爱的名小吃、品牌客商；对重点客商采用一对一招商跟踪，给予最大限度的条件优惠，确保进场，以形成对整体招商的示范效应。

4）带动个体进场。大客户是重点，个体是主体，从大客户入手，吸引个体满场；对个体客商制定切实可行的低门槛准入制；对个体客商采用广告推广，普遍招商，高举低打，逼客入场的销售策略。

旺场推广策略

在旺场推广策略上主要采取以下六种策略。

1）精确传播。锁定目标消费群体，针对性地选择传播渠道和宣传载体，尤其注重在区域内率先发动。

2）事件营销。在开业一段时间后，持续策划各类型契合目标对象的主题活动，包括体验参与性活动、表演展示性活动、展销促销性活动，活动要具备针对性、首创性、参与性、新闻性等特征。

3）内涵吸引。开业后要确保目标对象第一次逛商场有兴奋点，淘宝有得买，参与有得玩，美味有得吃，消费有得花，满足物质层面的消费需求、精神层面的自由需求。

4）随机消费。开业之初，必须下力气吸引客源，刺激消费者参观、体验、消费。其意义不仅在于充盈人气，更在于品牌传播和实现培育一批忠诚消费者。

5）双管齐下。一手抓上下九客源，一手抓区域目标客源。用人气带人气，以口碑促口碑，牢固确定“主动出击揽客，随机形象吸客”两手并举，确保开业后旺丁旺场。

6）雪球效应。必须给首次光临的目标对象留下深刻印象，使之产生强烈的认同感，乃至“逛不赢”的好感，通过他们喜欢成群结帮的消费习惯，滚雪球般地带动更多的目标对象前来消费。

Lianxianglou Restaurant

Founded in 1889, Lianxianglou Restaurant used to be a pastry shop and changed its name to "Lianxianglou" in Guangxu Period. In Qing Dynasty, grand secretary Chen Ruyue was fascinated by the pure lotus paste initiated by Liangxianglou and wrote down the restaurant title which has been used till this day. Liangxianglou features Chinese snacks and cakes. It is hypercritical in choosing material. Now it is a brand-name enterprise in China that assembles restaurant, food factory and food chains.

Lianxianglou is located at No.67, Dishifu Road, Liwan District. Its interior decoration takes lotus as theme. With wood furniture, Lianxianglou is of primitive simplicity and elegance.

莲香楼

莲香楼创立于1889年，最初是一家经营糕点美食的糕酥馆，后在清光绪年间改名为“连香楼”。清朝时期，大学士陈如岳被该楼首创的纯正莲蓉的清香风味所折服，挥笔题下“莲香楼”三个大字，楼名便一直沿用至今。莲香楼经营的是中式点心饼食，以莲蓉月饼为代表，品种有纯正的莲蓉月饼、榄仁莲蓉月饼和蛋黄、双黄、三黄、四黄莲蓉月饼等。其用料讲究，所选湘莲必是当年所产，制作出的馅料色泽金黄，油滑清香，月饼形状丰美。现在是集酒楼、食品工厂、贸易食品连锁店与一体的中国名牌企业，先后曾获得“国家特级酒家”“中华老字号”等称号，被誉为“莲蓉第一家”。

广州的莲香楼位于荔湾区第十甫路67号，厅堂内部布局以莲为主题雕梁画栋，有莲花形的吊灯、彩色投花玻璃和木椅圆桌等，古朴典雅。

Taotaoju

Taotaoju was founded in Guangxu Period of Qing Dynasty by Huang Chengbo. In 1927, it was taken over by Tan Jie, the Tea King. Taotaoju is one of the most famous teahouses in Guangzhou, mainly offering tea, refreshments and food and drinks.

Taotaoju located at Dishifu Road is a restaurant of Lingnan architecture style. Its interior is decorated as tradition Chinese style with calligraphy and painting of celebrities and elegant furnishings.

陶陶居

陶陶居始创于清光绪年间，原名“葡萄居”，创始人黄澄波。1927年，茶楼大王谭杰接手此店，招股集资重建新店，因新股东觉得原名“葡萄居”不雅而改名“陶陶居”，寓意于此品茗乐也陶陶。陶陶居是广州最有名气的茶楼之一，获得过“中华老字号”“国家特级酒家”等称号，主要经营名茶、茶点、茶食及酒菜餐饮等。

陶陶居现址在第十甫，是一座充满岭南建筑风格的酒楼，酒家大楼为三层钢筋混凝土结构，上盖有六角亭，墙面及亭中所选的彩画灰饰等极富岭南风格。厅房宽敞明亮，墙壁上挂有名人字画、诗词对联，陈设雅致，古色古香。